Designed for Dry Feet

Other titles of interest

WATER ENGINEERING

Flood Resistant Design and Construction, ASCE/SEI 24-05
2006. ISBN 0-7844-0818-1

Hydraulic Design of Flood Control Channels
1995. ISBN 0-7844-0067-9

Hydraulic Structures: Probabilistic Approaches to Maintenance
Walter O. Wunderlich
2005. ISBN 0-7844-0672-3

Soft Ground Improvement in Lowland and Other Environments
D. T. Bergado, L. R. Anderson, N. Miura, and A. S. Balasubramaniam
1996. ISBN 0-7844-0151-9

Soft Ground Technology
James L. Hanson and Ruud J. Termaat
2001. ISBN 0-7844-0552-2

Solutions to Coastal Disasters 2005
Louise Wallendorf, Lesley Ewing, Spencer Rogers, and Chris Jones (Editors)
2005. ISBN 0-7844-0774-6

Water Resources Engineering: Handbook of Essential Methods and Design
Anand Prakash
2004. ISBN 0-7844-0674-X

ENGINEERING HISTORY AND HERITAGE

America Transformed: Engineering Technology in the Nineteenth Century
Dean Herrin
2003. ISBN 0-7844-0529-8

Machu Picchu: A Civil Engineering Marvel
Kenneth R. Wright and Alfredo Valencia Zegarra
2000. ISBN 0-7844-0444-5

Tipon: Water Engineering Masterpiece of the Inca Empire
Kenneth R. Wright
2006. ISBN 0-7844-0851-3

Designed for Dry Feet
Flood Protection and Land Reclamation in the Netherlands

Robert J. Hoeksema, Ph.D.

Library of Congress Cataloging-in-Publication Data

Hoeksema, Robert.
Designed for dry feet : flood protection and land reclamation in the Netherlands / by Robert Hoeksema.
p. cm.
Includes bibliographical references and index.
ISBN 0-7844-0829-7
1. Flood control--Netherlands. 2. Reclamation of land--Netherlands. I. Title.

TC477.H64 2006
627'.409492--dc22

2005034784

Published by American Society of Civil Engineers
1801 Alexander Bell Drive
Reston, Virginia 20191
www.pubs.asce.org

About the cover

TOP: Levee along the lake, Ketelmeer at the mouth of the river, Ijssel; photograph by Martin Hendriks, used with permission.

BOTTOM, LEFT: Kinderdijk windmills used in an eighteenth century flood storage and detention project; photograph by the author.

BOTTOM, MIDDLE: The Maeslant Barrier shown in the closed position protecting Rotterdam harbor; photograph by Het Keringhuis, used with permission.

BOTTOM, RIGHT: Oosterschelde Barrier in action protecting the southwest delta from North Sea storm surges; photograph by Waterland Neeltje Jans, used with permission.

ISBN 0-7844-0829-7
Manufactured in the United States of America.

Contents

Preface

Few countries exist in which human activities have exerted a greater influence in shaping the landscape than the Netherlands. The purpose of this book is simply to tell a story—how this small country in Western Europe used and developed technologies over many centuries to create and maintain usable dry land in a very inhospitable environment. This story is often summarized by a Dutch saying, "God created the world, but the Dutch created Holland."

My interest in writing this book comes out of both my Dutch heritage and my interest in water resources. I am Dutch-American—I was born and raised in the United States, but all of my ancestors were Dutch. Even after earning graduate degrees in civil engineering hydraulics, hydrology, and water resources, and teaching courses on the subject, I never thought much about the water-related engineering achievements of my Dutch ancestors. This changed when a geographer colleague—Henk Aay—invited me to assist him in teaching an off-campus course. In 1995 we took 16 students to the Netherlands and spent four weeks exploring its rich history and amazing accomplishments in the area of flood protection and land reclamation. In 1996 I took a year-long sabbatical leave in the Netherlands. Even though my work focused on the subsurface (mapping soil layers), I took time to learn more about the history and technological developments. Henk Aay and I have since returned with students four times.

In the spring of 1998, I was invited to make a presentation on the subject of land reclamation and flood protection in the Netherlands to the Western Michigan Branch of ASCE. After making that presentation, I realized that many people knew that large parts of the Netherlands lie below sea level and are in danger of flooding. Many also knew that dikes and windmills were important tools in the struggle against water and that the Dutch reclaimed land from the sea. But few knew any of the details. How did the Dutch get to where all of this was necessary? What were the major accomplishments and the technologies used? What were some of the major projects, and what level of engineering effort was needed? The intent of this book is to simply fill in the details of this story.

The danger of living below sea level was made very real and this story was made more relevant with the recent flooding in New Orleans. In the case of the Netherlands, 65 percent of the country is situated below the level of high tide. A significant portion of the 16 million people who live in the Netherlands would be potential victims from a storm surge disaster like that generated by hurricane Katrina. Maintaining an acceptable level of safety is a critical, ongoing task.

In conversations with Dutch people about flood protection or land reclamation technologies, they will often refer to the process of maintaining dry ground as "keeping their feet dry." This phrase was the inspiration for the title of this book. The first chapter tells the entire story of the history of Dutch flood protection and land reclamation from dwelling mounds constructed as early as 500 BCE to large storm surge barriers completed at the end of the twentieth century. Chapter 1 also provides the social-political-economic context for these developments. Chapters 2 through 8 expand on the details first presented in Chapter 1. These chapters are organized by era, project, and/or geographic/physical setting. Chapter 9 covers water resources management.

Henk Aay and I teach an excursion-oriented course. We provide background through lectures, but much of the learning comes on the excursions that the students take throughout the course. In addition to providing historic and technical background, I want the reader to be able to use this book as an excursion guide on a trip to the Netherlands. To make this possible (without turning it into a typical travel guide), I have included, at the end of many of the chapters, a list of "Places to Visit." Furthermore, Chapter 10 outlines six possible excursion routes that one might follow to see many of these places. If you have the opportunity to follow one of these excursions, I would appreciate feedback for further improvement. Please send any comments to rhoeksem@calvin.edu.

This book is intended for English speakers who cannot read Dutch. There are many Dutch language resources on this subject, but English language resources are limited. The book is intended to pique the interest of civil engineers, but it is easily readable even by those without any technical training. If, after reading this book, you are interested in learning more about the topic, the best English language book available is *Man-Made Lowlands* written by Gerard van de Ven (van de Ven 2004).

Finally, a comment is in order about the use of Dutch words in this book. The names of many of the places and physical/geographic features could be presented in either English or Dutch. I decided to consistently use Dutch with the exception of the two most well-known features—the North Sea and the Rhine River. All other names will be given in Dutch (such as Haarlemmermeer instead of Lake Haarlem and Noord-Holland instead of North Holland).

Robert J. Hoeksema

Acknowledgments

Many people have helped me with the preparation of this book. I am, first of all, grateful for the assistance of those who spent many hours reviewing the text. Their comments and insights greatly improved the quality of this publication. In particular, I am grateful to the following people, who reviewed the entire manuscript for technical detail, readability, grammar, and style: Henk Meijer, Dutch geographer and former director of the Information and Documentation Center for the Geography of the Netherlands; Claire Schwartz, PE, Senior Associate at Fishbeck, Thompson, Carr, and Huber; Laura Hoeksema, editorial production assistant at Omegatype Typography; and the two technical reviewers selected by ASCE Press. I also thank the following four persons who reviewed individual chapters because of their expertise in particular areas: Frans van Liere, Professor of History at Calvin College; Pieter Kooi, Assistant Professor of Regional Archaeology of the Northern Netherlands at the University of Groningen; Jan Verbruggen, author of the book *Steam Drainage in the Netherlands*; and Sybe Schaap, Chairman of the Waterboard Groot Salland and Chairman of the Union of Waterboards. I am especially grateful that I was able to meet and work with Jan Verbruggen, a noted expert in steam drainage, before his death in May 2005.

I also received help in obtaining figures and photographs for use in this book from several individuals and organizations. The following organizations deserve recognition: The Royal Dutch Geographical Society, The Groningen Institute of Archaeology, The Dutch Windmill Society, The Cruquius Museum, The Rotterdam Municipal Archive, The Social Historical Center for Flevoland, The Royal Archive in Flevoland, AVIODROM Aerial Photography, Rijkswaterstaat Construction Services, Waterboard Groot Salland, Waterland Neeltje Jans, The Netherlands Photography Museum, Rijkswaterstaat Directorate Zuid-Holland, and The Keringhuis. I am also very grateful to Gerard van de Ven for allowing me to use several figures from his book *Man-Made Lowlands*.

I also thank those who have encouraged me throughout the writing of this book. Two persons deserve special mention. My wife, Brenda, provided support throughout this process. I am especially grateful to her for making it possible for me to travel to the Netherlands on a regular basis. She lovingly handled all household and family responsibilities during the month-long trips with students. I also want to thank my geographer colleague Henk Aay (see Preface). Henk is responsible for initiating my interest in the topic of this book. Over many years of working with Henk, I have learned much about Dutch history, geography, and society. He and I have spent countless hours and driven thousands of kilometers traveling throughout the Netherlands. With his interest in the environment and geography of the Netherlands and my interest in its water-related technology, we are constantly learning from each other. He is a good friend and colleague.

Finally, I am appreciative to Calvin College for providing financial support for this project by granting me a sabbatical leave.

1

Land Reclamation and Flood Protection in the Netherlands: An Historical Overview

The Netherlands is a small country in Western Europe that has faced unique challenges in the area of water control and management. The northern and western part of the country is a flat, low-lying region situated in the delta of three major European Rivers—the Rhine (in Dutch, Rijn), Maas, and Schelde. Over the past 2,000 years the consequences of human habitation and natural events have combined to make life in this region challenging. To use low-lying land for agriculture required removal of excess water by the construction of field drains and ditches. Once dry, the clay and peat soils subsided, requiring further deepening of the drainage network. This began an irreversible pattern of land subsidence that continues today. To make matters worse, a large volume of peat soil was removed for use as fuel. Coinciding with this loss and subsidence of the land was a period of postglacial sea level rise. These factors, along with the natural action of tides and storm surges, have created a landscape that is constantly changing. Today about one-quarter of the Dutch land lies below mean sea level. [This is based on 34,000 square kilometers (13,000 square miles) of dry land including water courses less than 6 meters (20 feet) wide.] Without the benefit of dunes, dikes, canals, and pumps, 65 percent of the land would be under water at high tide. Without the river dikes, even more land would be flooded on a regular basis.

The Dutch people have thrived in this environment because they were willing to fight to "keep their feet dry." This fight has involved the use of a variety of different technologies for drainage (using both gravity and mechanical means), for flood protection, and for reclamation of land. This fight would never have been won without an organized effort and without the creative work of Dutch engineers.

This chapter covers the entire history of flood protection and land reclamation from the beginning of refuge-mound construction to present day movable gates and bladder dams. It touches on the major projects, leaving details to later chapters. This chapter also includes the political, economic, and social history of the region putting the flood protection and land reclamation activities in a broader historical context.

Today's Dutch lowland landscape consists of hundreds of water management units called *polders*. A polder is an area completely isolated from the surrounding area, usually by a berm, dike, or embankment, in which the groundwater levels are artificially controlled. This control involves a variety of drainage schemes—usually ditches and canals that deliver the water to the perimeter of the polder. The water is then released from the polder using a pump or sluice. A pump is needed if the water outside the polder is always higher than the water inside the polder. If the water outside the polder is occasionally lower than the water inside, then a sluice can be used.

Figure 1-1: The polder concept. Water is drained from low-lying areas by a network of drainage ditches, pumps, sluices, and canals. Typical elevations (above and below sea level) are shown in fine print.

Reprinted, by permission of the Royal Dutch Geographic Society, from IDG 1994.

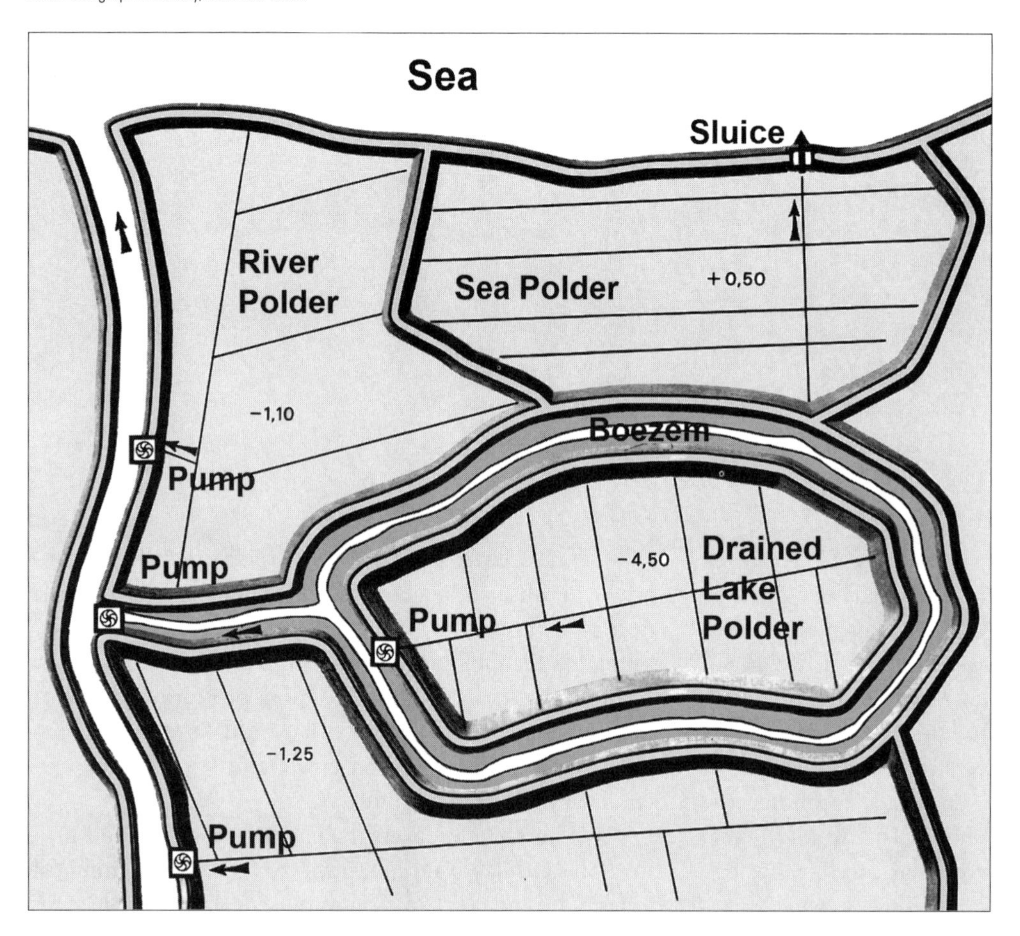

A sluice is a gate or valve that allows water to flow out of the polder but not back in. Figure 1-1 shows several types of polders. A polder can lie along a river or the sea for easy drainage. Some polders—like drained lakes—lie further inland and require special drainage routes to get the water to the sea. These polders also require storage areas, called *boezems*, to store water during high tides or periods of extended high receiving water levels. One goal of this book is to discover how a large part of the Netherlands was transformed over the past 2,000 years into this polder landscape.

The Dutch Landscape— Geography, Geology, and Climate

To properly set the stage for the developments in this book, an understanding of the geographic setting, geologic conditions, climate, and hydrology is needed. A significant part of the story revolves around issues of soft organic soils that tend to subside, a low flat topography that is susceptible to storm surges, a network of river branches that frequently flood, and wind-driven erosion that increases the size of inland lakes and damages coastal defenses.

Figure 1-2: The Netherlands and neighboring countries, its provinces, and several major cities.

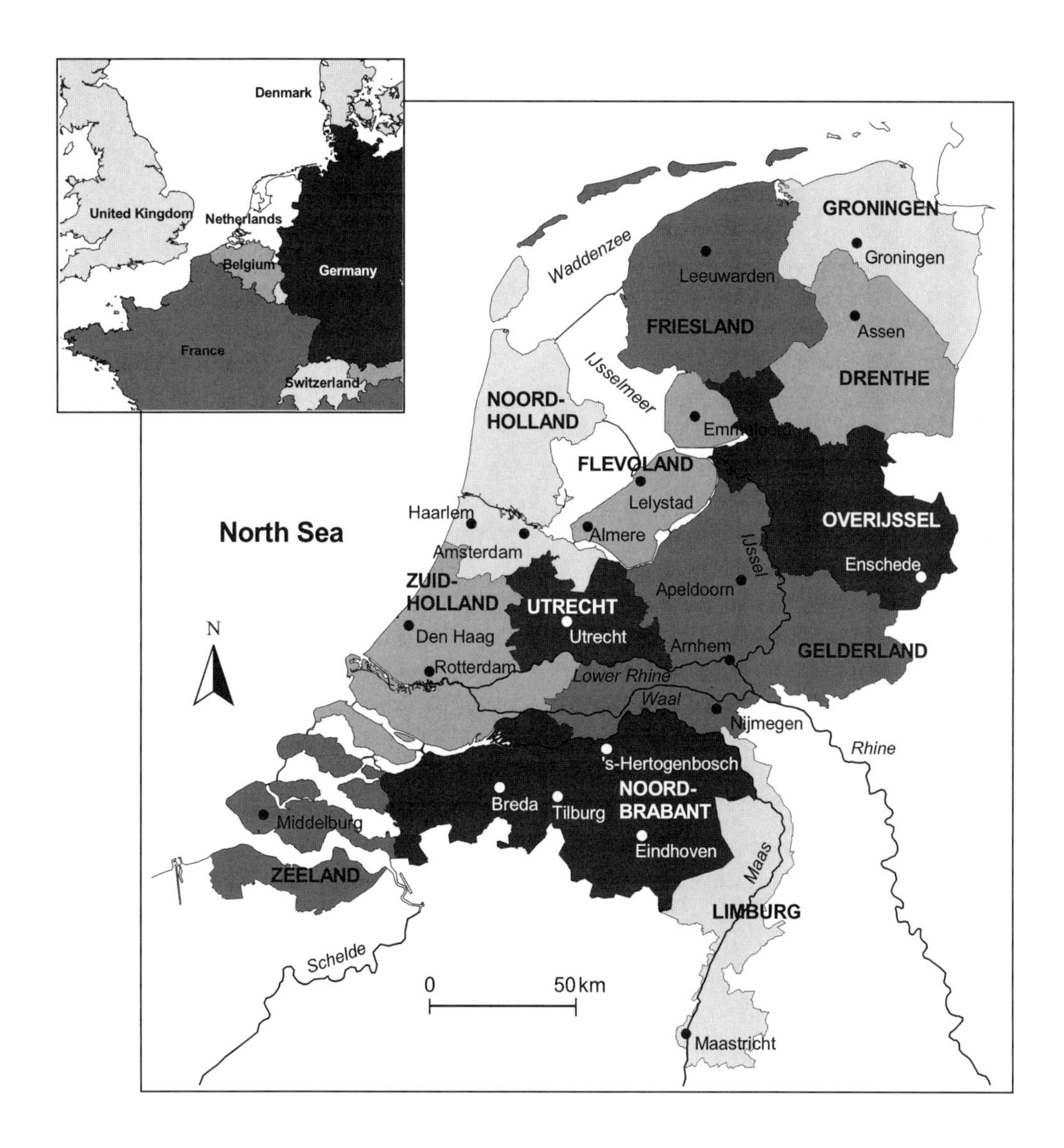

Geography

The inset map in Figure 1-2 shows the location of the Netherlands situated with Germany to the east, France and Belgium to the south, and the United Kingdom to the west. The size of the Netherlands is 41,500 square kilometers (16,000 square miles). This area includes all of the lakes, estuaries, and territorial waters. The total land incorporated in municipalities is 37,000 square kilometers (14,000 square miles). If all water bodies larger than 6 meters (20 feet) wide were eliminated, the total dry land would be 34,000 square kilometers (13,100 square miles) (Henk Meijer, personal communication). This is somewhat larger than the state of Maryland at 12,400 square miles. Figure 1-2 also shows a more detailed map of the country, highlighting the twelve provinces, major cities, and significant water bodies. The 2001 census counted 16 million inhabitants. Around 45 percent of the population lives in the three western provinces—Noord-Holland, Zuid-Holland, and Zeeland (about 20 percent of the land area). These provinces

▾ **Figure 1-3:** Current topographic contours of the Netherlands.
Redrawn from SWAVN 1986, Fig. 40.

▸ **Figure 1-4:** A farm house along the Hollandse IJssel River near the lowest point in the country at –6.74 meters NAP.

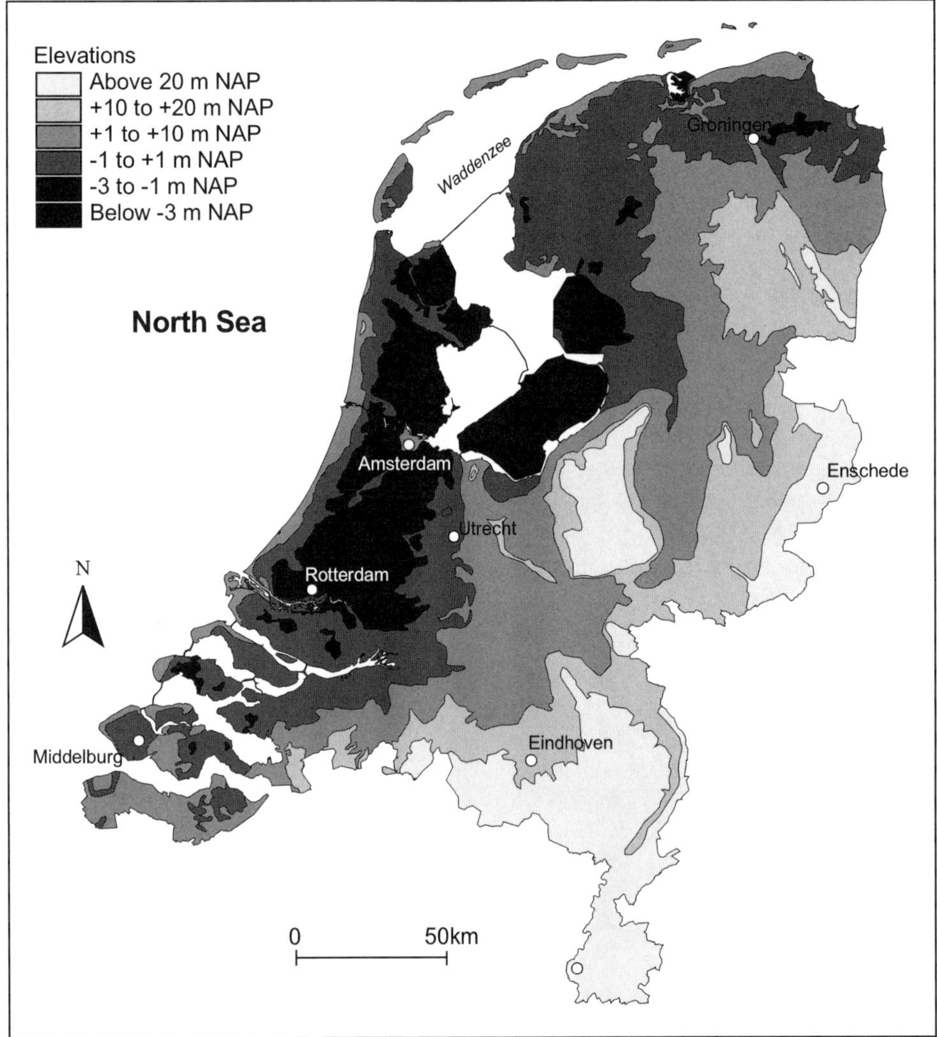

are also some of the most threatened by flooding from the sea. With the exception of Flevoland (100 percent below sea level), these provinces have the most significant land area below sea level.

It is well known that a large portion of the Netherlands lies below sea level. **Figure 1-3** is a contour map showing ground surface elevations. The datum from which all elevation measurements are made is the "Normaal Amsterdams Peil" or Amsterdam Ordnance Datum—better known as NAP. The NAP is the average sea level as measured in the Amsterdam harbor at a time when it was still in direct contact with the sea. This reference level was first established as the Amsterdams Peil (AP) in 1682. It was later revised based on new measurements made in the nineteenth century. The lowest elevation of 6.74 meters (22.3 feet) below NAP occurs at a location just north of Rotterdam. **Figure 1-4** shows a home located along the Hollandse IJssel River near this low spot.

Figure 1-5: Current soils of the Netherlands.
Redrawn from SWAVN 1987, Fig. 18.

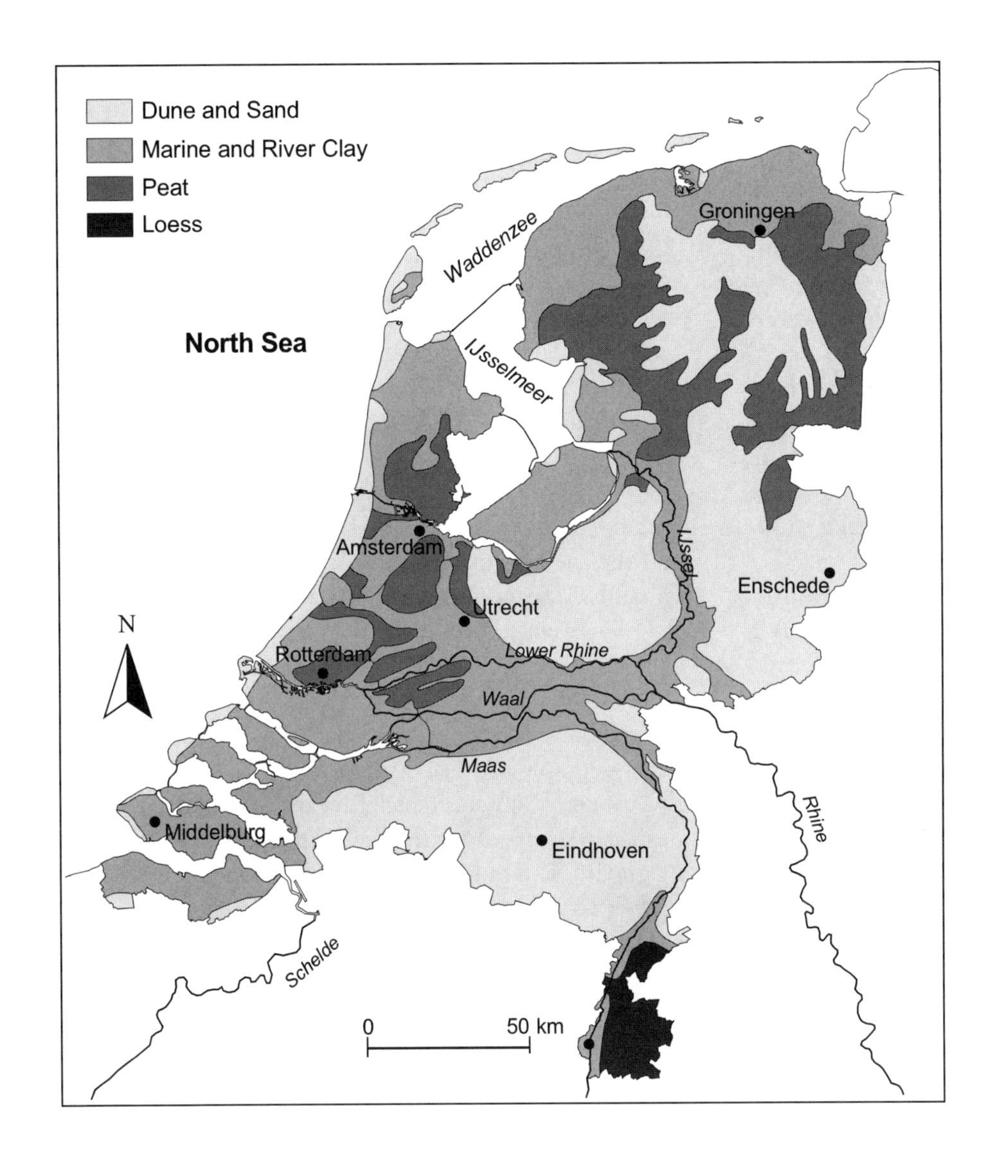

Geology

At the surface, most of the soils in the Netherlands were deposited in the Pleistocene and Holocene epochs. **Figure 1-5** is a map showing the current surface soil landscapes of the Netherlands. The lower elevations in the north and west are dominated by peat, marine clay, and river clay. The peat and clay are significant because of their tendency to decay (peat) and consolidate, resulting in land subsidence. The clay was an important element in dike building. Early in the Pleistocene period much of what is now the Netherlands was submerged. Sediments deposited by the sea and the rivers built up the land. These river sediments were primarily sand and gravel originating from the Alps and other central European mountain ranges. The Pleistocene period was characterized by a number of ice ages. The main effects of the ice ages in the Netherlands during this period were the variations in sea level and the plowing action of the ice sheets. During the periods of ice advance, the sea level was low and the rivers were actively down cutting through the sand and gravel. During the warmer periods, the sea level rose and the rivers responded by delta building. Only one of the ice advances recognized in Europe actually reached the Netherlands. The lobes extended to a line between Haarlem and Nijmegen. As a result, the great rivers that ran north through this region (Rhine and Maas) were diverted to the west (see the map in **Figure 1-5**). In addition, the plowing action of the ice sheets created ice-pushed ridges or moraines, which reached heights between 50 and 100 meters (164 and 328 feet) in the provinces of Utrecht, Overijssel, and Gelderland. The ice also deposited vast amounts of boulder clay, up to 30 meters (98 feet) thick, in Drenthe and southern portions of Friesland. This clay was used for dike building in the twentieth century.

The Holocene period saw additional deposits placed on top of the existing Pleistocene. This was a warmer period characterized by sea level rise, peat formation, and development of coastal dunes. Fluvial deposits continued along the great rivers.

Peat is formed by the accumulation of organic material that has not decomposed. Peat forms readily in temperate, moist regions. It is aided by high groundwater levels. The rise in sea level during the Holocene epoch resulted in a lower gradient in the lower reaches of the rivers. This created large areas of stagnant water and provided a good environment for peat growth. Areas along the coast turned into shallow lagoons, providing sites for peat growth. Furthermore, the rising sea level produced higher groundwater levels near the coast, resulting in further accumulation of peat.

As the sea level rose, the coast continued to move eastward. The coastline consisted of a sandy coastal zone, mud flats, salt marshes, and peat. Around 3000 BCE the rate of sea level rise started to decrease. This allowed the coastal dunes—commonly referred to as the "old dunes" —to form. This closed off the area of mud flats and salt marshes, resulting in accelerated peat growth. Near the coast, marine clay was deposited on top of the peat by the rising sea. In the north this started around 1000 BCE and in the south after 300 BCE. Elsewhere, peat growth continued unhindered until around 800 CE. After that time growth eventually came to a halt only through human intervention. **Figure 1-6** shows the extent of the peat deposits around 800 CE.

After 1000 CE coastal erosion steepened the sea bottom. The sand that was released in this process created a new set of coastal dunes several kilometers wide. These are referred to now as the "young dunes."

Climate and Hydrology

The climate in the Netherlands is strongly influenced by the North Sea and the Gulf Stream. Despite being situated at 52° north latitude, the annual mean temperature in the center of the country is about 10° Celsius (50° Fahrenheit). Along the coast the average January temperature is 2.6° Celsius (36.7° Fahrenheit) and the average July temperature is 16.2° Celsius (61.2° Fahrenheit). Along the coast the average wind velocity is 5.8 meters per second (28 miles per hour) and the annual precipitation depth is 734 millimeters (28.9 inches) (NHV 1998).

The hydrology is strongly influenced by the low, flat topography. Surface water in the low western part of the country requires artificial means to drain to the sea. Surface drainage from the higher parts of the country is by gravity. The most significant component of the water budget for the country is the flow from the two major rivers, the Rhine and the Maas. In an average year they provide 71 percent of the water (the other components being 26 percent precipitation and 3 percent other rivers) (NHV 1998).

The history of flood protection and land reclamation is closely tied to the continual rise in the sea level. Discussions about sustainability in the future are closely linked to the expected rate of sea level rise. Over the past 2,000 years there have been several periods where the sea level rose at a faster pace. These marine transgressions were periods of stormier weather and heavier precipitation, resulting in more frequent flooding. Three sea-transgression periods have been recognized—the late Roman (200 CE to 500 CE), the early medieval (800 CE to 950 CE), and the late medieval (1130 CE to 1500 CE) (Lambert 1971). In this last period a large number of storm surges occurred, causing great damage and loss of life.

Figure 1-6: Soils and coastline around 800 CE with a comparison to the current coastline.
Redrawn from SWAVN 1984, Fig. 5.

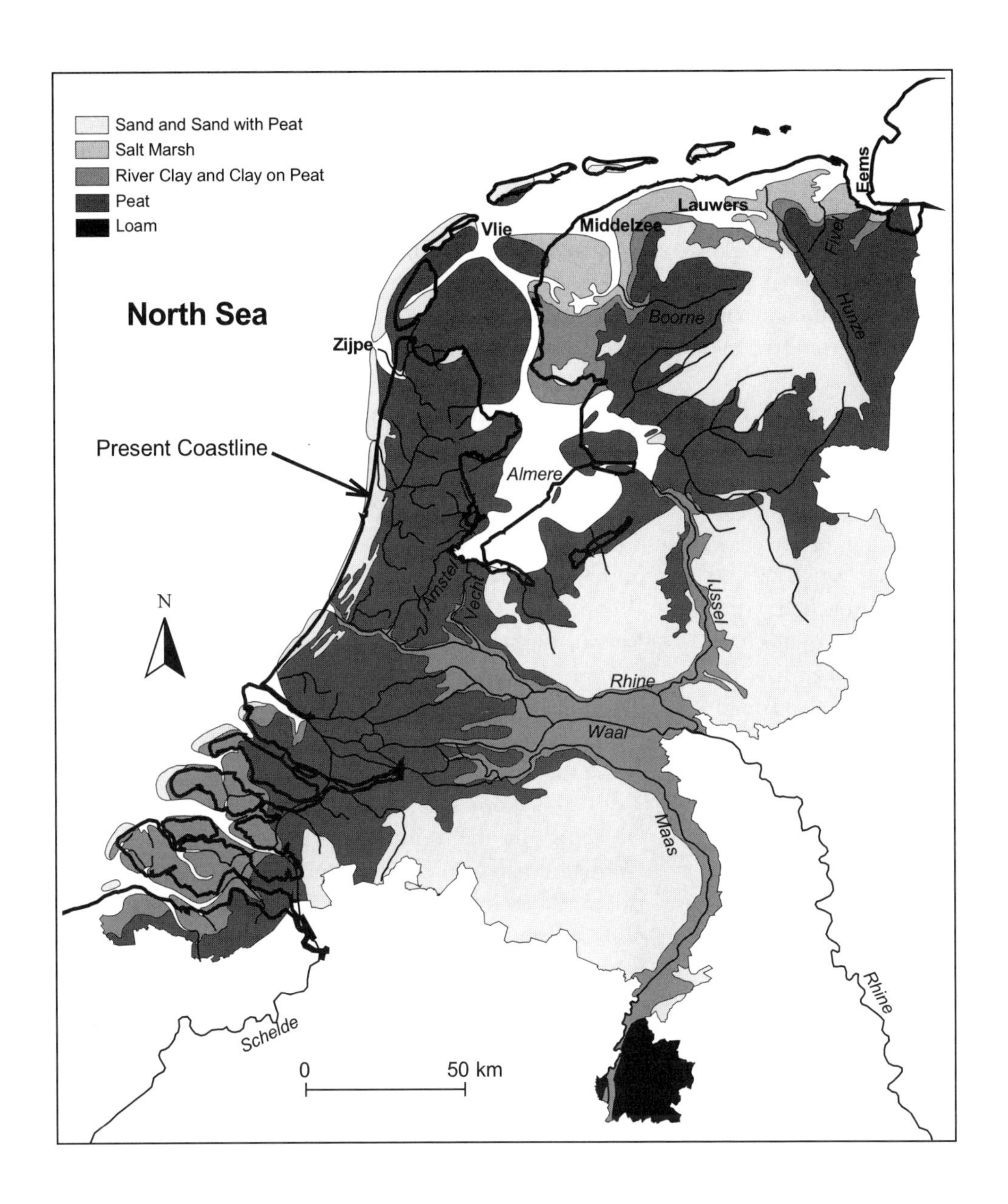

Early Settlements through the Roman Period

The first inhabitants of the Dutch lowlands to move earth to create protection from the sea were the earliest inhabitants of the northern regions (now the provinces of Groningen and Friesland). These people were the first to make significant, recognizable changes to the living environment. We might consider them to be the first Dutch hydraulic engineers—the first to construct ways to keep their feet dry.

In the first millennium BCE most of the northern inhabitants resided on the high, 5 to 25 meters (16 to 82 feet) above sea level, sandy glacial deposits of the Drenthe plateau. In search of better pasture for their livestock, some of these people began to migrate toward the coast beginning around 500 BCE. The coastal region consisted of clay deposits that had been laid down since the last glaciation. Coarse particles were deposited near the coast while the finer particles settled further inland. The coarser sediment tended to subside less, creating slightly elevated salt marsh banks. During periods of marine regression, these people migrated toward the coast and settled on these salt marsh banks, which were often less than a meter above normal high tide.

These salt marsh banks did not provide much protection during storms. Storm surges could easily inundate their settlements. In response to this threat, the people who settled in this region began to build the earliest form of protection from flooding. They used domestic and animal waste to create small refuge mounds. When the floodwaters approached, they retreated to these mounds for safety. These mounds have been given several different names. In the region of the Dutch province of Friesland they are called *terpen* (singular *terp*). Further north in Groningen they are called *wierden* (singular *wierde*). These mounds can also be found throughout the Holocene coastal deposits from Belgium to Denmark. Chapter 2 covers the development of dwelling mounds as a form of flood protection and the resulting impact on the social and economic character of the northern coastal region.

From 57 BCE to 406 CE the southern part of the Netherlands was part of the Roman Empire. The Romans were not interested in the Dutch lowlands region for its physical beauty, natural resources, or inviting climate. It was a swampy, damp, cold region with little or no mineral resources available and poor agriculture. The primary reason for Roman settlement was its strategic location at the mouths of the Rhine and the Maas Rivers. Occupation along the Rhine (which at that time emptied into the North Sea near where the city of Leiden is located today) stabilized the frontier. Fortifications along the Rhine were built to establish a base for defense against northern Germanic tribes. The Roman influence was more significant south of the Rhine than north.

The official beginning of Roman rule occurred in 57 BCE when Julius Caesar declared the Rhine River the Roman frontier. The Romans were the first to make hydraulic modification to the Dutch rivers. In 12 BCE a Roman military commander, Drusus, in exchange for labor and a release from taxation, improved some of the water links from the Rhine to the north. He ordered a dam to be built to improve navigation to the north. He also dug a canal between the Rhine and the Vecht Rivers. In 45 CE the Roman commander Corbulo constructed a canal that connected the mouths of the Rhine and the Maas Rivers.

The Germanic people that occupied the region along the Rhine River at this time were known as the Bataves. Under Roman rule the Bataves settled along the natural river levees that follow the Rhine and the Maas. These locations were artificially raised with the debris of human occupation, creating dwelling mounds along the Rhine. The Bataves were not incorporated into the Roman Empire but were considered to be Roman allies.

Excavations at Utrecht and Valkenburg (near Leiden) indicate that even the Romans built refuge mounds. A Roman fort at Utrecht was built 6 meters (20 feet) above the surrounding marshes.

The Roman presence in the Dutch territories weakened in the third century CE when raids from Germanic tribes forced the Romans to retreat. Roman rule was reestablished in the fourth century, but around 400 CE Roman military and political control in Western Europe faded and, eventually, disappeared.

Middle Ages

The Middle Ages (roughly the fifth through the fifteenth centuries CE) was a period of tremendous change in the region that is now the Netherlands. Political authority became highly decentralized in the transition from Roman and Frankish rule to the feudal system. As the population in the Dutch lowland area increased, large areas of peat swamp were "reclaimed" for agricultural use. Subsequently, many of these areas were then lost as a direct result of increased settlement and frequent flooding. Some of the earliest technologies to again reclaim these flooded lands were implemented in this period. Note that the word "reclaim" refers to both the initial development of land for agricultural use and the subsequent redevelopment after these lands were flooded.

Historical Developments

At the end of Roman rule, control of the Dutch region was split three ways. The Saxons ruled in the east, the Frisians occupied the north, and the Franks were in charge south of the great rivers. As the Romans retreated, the Franks continued to gain land and eventually became the dominant power. In the eighth century the Frisians were annexed by the Frankish empire, extending it to the north. In 785 Frankish rule was extended even further when Charlemagne defeated the Saxons. Along with the expansion of the Frankish Empire came the spread of Christianity throughout Western Europe. In 800 the Pope crowned the Frankish king, Charlemagne, Emperor of the Romans, in an attempt to recreate a Roman, that is Christian, empire.

Under the rule of Charlemagne local regions (counties) were administered by counts, trade flourished, and—by the ninth century—the region had become prosperous. After his death in 814, Charlemagne's empire was split several times, bringing disorder to Western Europe. The feudal system that characterized the Middle Ages was a way for the people to have stability and protection during a time when a strong central authority did not exist. Political and economic control became highly decentralized.

The period between 850 and 1000 was marked by frequent raids by Vikings. The inhabitants of the Dutch region spent more time on defense and less on agricultural production. In addition, climate changes made farming more difficult, and the region fell into a less prosperous period.

By the eleventh century all of the administrative power in the region was in the hands of local counts, dukes, and bishops who were subject, in name only, to the Holy Roman emperor. Their feudal principalities became the basis for the provinces that have played an important role in Dutch history. In the eleventh and twelfth centuries an increase in population, trade, and industry resulted in a resurgence of the importance of the towns. Towns with access to the North Sea grew in importance, as the larger trading vessels could not navigate the inland waters. By the fifteenth century many of the towns operated as independent communities.

In the fifteenth century, Philip the Good used marriage, inheritance, and war to unite the Dutch provinces under the control of the dukes of Burgundy. In 1464, he created the first States-General, a gathering of representatives from all of the provincial councils—a precursor to the current Dutch parliament. The dukes of Burgundy administered the southern provinces directly, while the other provinces were placed in the hands of appointees, known as *stadholders* (in Dutch, *stadhouder*). Thus, central power was consolidated in the hands of a single family. With Philip's son, Charles the Bold, the male line of the Burgundian dynasty ended. The marriage of his daughter, Mary, to Maximilian, the son of Emperor Frederick III of the House of Hapsburg, eventually placed the Dutch provinces under Hapsburg rule.

In 1500 Charles V was born. He was the grandson of Maximilian—now Holy Roman Emperor and German King. His maternal grandfather was Ferdinand II of Spain. Charles V began his rule of the Dutch provinces in 1506. When Ferdinand died in 1516, Charles became the King of Spain. When Maximilian died in 1519, Charles was elected Holy Roman Emperor and brought Hapsburg power to its greatest level. Under Charles V, the former Burgundian Netherlands (including the current countries of Belgium and Luxemburg) were united under the title of the "Seventeen United Netherlands."

Changes in the Dutch Landscape

The Middle Ages was characterized by increased settlements in the low-lying areas and by human activities, which significantly changed the landscape. This was also a period of marine transgressions, which, combined with human activities, increased the frequency of flooding, resulting in significant changes to the Dutch coastline. This was a period in which those who lived in this region generally lost the battle against the sea, despite the implementation of some of the earliest land reclamation and flood protection technologies.

Until 800 the land that is now the Netherlands was shaped exclusively by natural forces, primarily peat formation, sedimentation, and erosion. This process has been described in the earlier parts of this chapter. **Figure 1-6** shows the configuration of this region around 800. The early medieval coastline was quite different from today's (note present coastal outline in **Figure 1-6**). The regions of particular interest in this book are the vast peat areas (compare to **Figure 1-5**), the old dune coastline, the tidal salt marsh in the north, the region of the rivers, and the peat and clay on peat delta of the southwest. Around 800 the Dutch coastline was almost completely closed. A few openings existed—the large rivers Rhine, Maas, and Schelde drained to the sea through the coast. In the northeast much of the drainage was through the rivers Fivel, Hunze, and Eems. A number of sea arms penetrated the coast, such as the Zijpe, the Vlie, the Middelzee, the Lauwers, and the Eems.

Most of the peat was sphagnum peat or blanket bog formed by the decomposition of sphagnum species. This type of peat, which forms above the groundwater level, gets its water supply from precipitation. Near the coast peat domes formed with diameters of 10 to 15 kilometers (6.2 to 9.3 miles) and elevations of 3 to 4 meters (9.8 to 13.1 feet) above NAP. Inland the sphagnum peat rose to elevations as high as 3 meters (10 feet) above NAP. Within this extensive blanket bog there were a series of lakes and marshes called the Almere. The surface of the Almere was 10 to 20 centimeters (4 to 8 inches) above NAP and drained to the sea via the Vlie (see **Figure 1-6**). In some locations the peat had been overlain by clay or sand deposited by the North Sea and the large rivers.

When the Viking raids stopped at the beginning of the eleventh century, a period of marine regression started, and the climate improved. As the population increased, the inhabitants looked to the peat swamps for new agricultural land. The process of transforming or reclaiming peat swamps into useful land was rather simple. Digging ditches in the peat bog provided enough drainage to allow productive use of the land. The drainage ditches were dug along the edges of the parcels. The ditches generally ran perpendicular to the ground surface contours and drained into natural streams or larger drainage canals. As a result, a number of parcelization patterns developed. Feather and fan parcelization patterns can be found along natural streams draining peat domes (see **Figure 1-7**). Parallel patterns were common along drainage canals.

Along the coast the peat reclamation process started from the dunes and proceeded inland. The farmers built their villages along the dunes and set up long, narrow parcels, allowing the farmsteads to be located near the villages. At the back of the parcel, away from the farmsteads and at the upper elevations, the farmers built a ditch and dike to keep the water from the undrained areas from flowing through their drained areas.

Drainage of the peat allowed the inhabitants to create usable agricultural land, but the drawback was subsequent land subsidence. The three major causes of this subsidence were soil shrinkage, consolidation, and oxidation. Shrinkage is the change in soil volume produced by capillary stress during the drying of a soil. It is a function of the initial water content of the soil and the loss of water by evaporation and transpiration. Consolidation is the gradual compression of a cohesive soil due to an increase in the effective stress at some depth. When the groundwater level is lowered, the buoyancy forces at any depth in the soil are reduced. As a result there is an increase in the soil contact pressure or effective stress and a compression of the soil layer. The amount of time for the compression to become effective is controlled by the rate at which the water squeezes out of the soil. The time-delayed compression of soft soils, like clay and peat, is referred to as consolidation. Consolidation and shrinkage caused some of the land subsidence. But a more significant cause was oxidation or decay.

Peat is plant material that has been deprived of oxygen, stopping the action of decay-causing microorganisms. As the peat dried and was plowed, the decay resumed. The addition of manure to improve the soil provided new microorganisms to stimulate further decay and oxidation. The loss of water also allowed the soil to reach higher temperatures, further accelerating decay and significantly increasing the land subsidence. It is estimated that decay alone caused up to 85 percent of the land subsidence (Ritzema 1994). In some areas 3 to 4 meters (10 to 13 feet) of peat disappeared completely. By the seventeenth century the peat domes had dropped to below NAP. As the reclaimed land subsided, high groundwater levels made the land unusable. Thus, the reclamation process was pushed further up the peat bog as is shown in **Figure 1-8**. In many cases the new

▾ **Figure 1-7:** Fan and feather parcelization schemes in the peat regions.
Redrawn from van de Ven 2004, Ch. 2, Fig. 14.

◂ **Figure 1-8:** Progressive reclamation in the peat bogs and accompanying land subsidence.
Redrawn from van de Ven 2004, Ch. 2, Fig. 21.

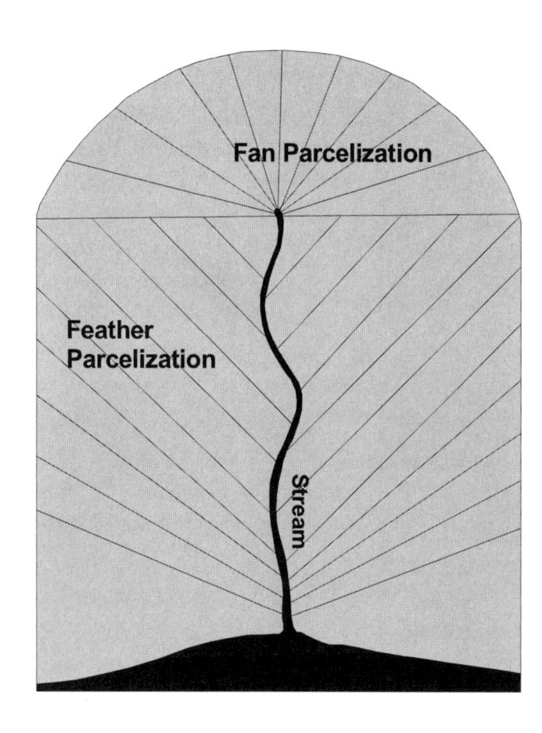

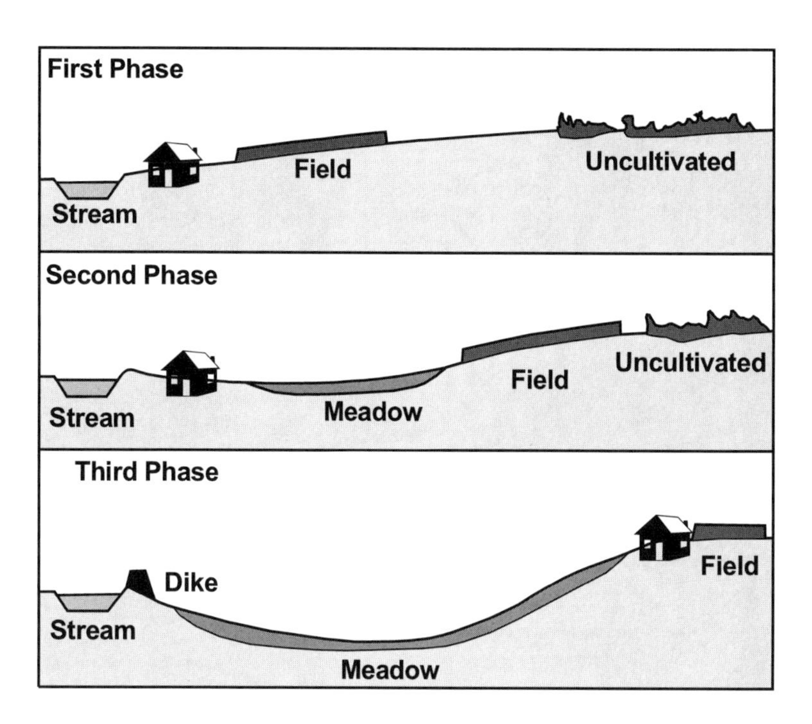

reclaimed parcels simply followed the existing drainage patterns and became extensions of the old parcels. This process was repeated several times and often required the relocation of the entire farm.

This process can still be recognized in the landscape today. **Figure 1-9** is a current map showing the area around the town of Nieuw-Loosdrecht. Here the feather, fan, and parallel parcelization can still be seen. The fan-shaped parcels surround a former peat dome. The areas that were first drained follow the Vecht River along the left side of the map. These areas eventually became completely flooded. The removal of peat for fuel exacerbated the problem.

The first water management organizations were instituted in the twelfth century. The counts and bishops, who had all of the administrative power at the time, made contracts with local communities to embark on various reclamation projects. These contracts gave the community or village the authority to manage and maintain the project. Since everyone in the village shared the benefits of reclamation, everyone also shared in the construction and maintenance of the drains and ditches. As the projects became more complex and began to involve more villages, the local authorities established regional water administrations called water boards or *waterschappen* to manage the parts of the system that affect multiple villages. Each village appointed members to these boards. So, the earliest form of water management, established in the twelfth century, kept control at the local level. The first water board was the Lekdijk Bovendams water board established in 1122. Water boards were established for many purposes. These include operation and maintenance of a particular dam or sluice or a system of dikes and drainage canals. As part of their work, they set up rules of operation and provided inspection. They also acquired legislative, judicial, and executive powers. Today, water boards continue to manage water resources at local and regional levels. Chapter 9 covers the history of the Dutch water boards.

From the middle of the twelfth century many storm surge related floods were recorded. The following is a list of several of these (van de Ven 2004):

- **1134:** A storm surge in the southwest increased the size of the sea arms. Extensive diking to protect the land followed this storm event.

▾ **Figure 1-9:** Results of parcelization in the peat region near Nieuw-Loosdrecht.

▾ **Figure 1-10:** Noord-Holland lakes and the Zuiderzee in 1250.

Redrawn from SWAVN 1984, Fig. 5, and van Duin and de Kaste 1990, Fig. 7.

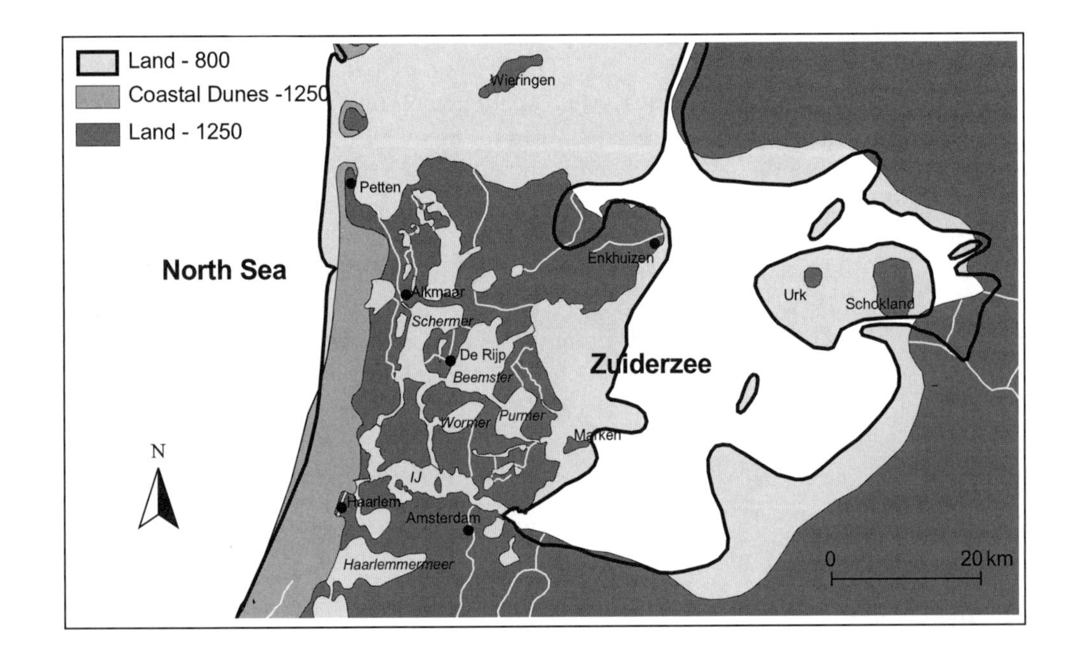

- **1163:** A storm surge occurred that resulted in the complete closure of the mouth of the Rhine River and the redistribution of Rhine River water to other outlets. The section of river leading to the closed mouth is now referred to as the Old Rhine (Oude Rijn in Dutch).
- **1170 and 1196:** Storm surges opened up Lake Almere to tidal influence. The lake became a sea arm and eventually was called the Zuiderzee.
- **1214, 1219, and 1248:** Storm surges expanded the Zuiderzee and Noord-Holland lakes.
- **1287 and 1288:** These storm surges greatly expanded the size of the Westerschelde in the far southwest. Dikes along the Eems in the far northeast burst, causing the creation of the Dollard sea arm.
- **1373:** A storm surge destroyed the area known as the Riederwaard in the southwest and created a sea arm along the Westerschelde, the Braakman.
- **1421:** A storm surge, known as the Saint Elisabeth's Day Flood, destroyed the area east of Dordrecht known as the Grote Waard. A large part of this area is still unreclaimed today. The destruction of the Grote Waard was probably not a single catastrophic event. It is believed that the loss came about incrementally as political complications hindered the repair of many dikes. After some time 28 villages were simply abandoned. Visitors to Dordrecht in 1500 recalled seeing a large inland sea with church and castle towers still rising above the water. This picture of a drowned land led to many of the stories told today about a great disaster with thousands of casualties.
- **1472 and 1509:** Floods in these years contributed to the creation of the Haarlemmermeer (Lake Haarlem).
- **1532:** A storm destroyed the town of Reimerswaal and the surrounding land on the southwest island of Zuid Beveland.
- **1570:** The All Saints Flood was one of the largest flood disasters to occur in this period. It affected the entire coastal region.

Tremendous change occurred in the Dutch landscape in the Middle Ages especially after 1000. The next sections will focus on individual regions.

Developments in Noord-Holland

The region west of the Vlie (see **Figure 1-6**) is now part of the province of Noord-Holland north of Amsterdam. This area experienced some of the greatest changes in the Middle Ages. Here the peat was drained in parallel ditches toward the small rivers. The area around the rivers subsided first. The rivers grew into lakes. Wave action on the lakes eroded the shores, causing them to grow out of control. The lakes, such as the Schermer and the Beemster, were eventually named after the rivers from which they were created. Many of these lakes were formed between 1150 and 1250.

The region west of the Vlie was not only threatened by the expanding lakes but it was also threatened by the sea and the growing lake Almere. Erosion from storm surges in the twelfth century made permanent changes to the mud flat area in the north. These storms also greatly expanded Almere, turning it into the sea arm later to be renamed the Zuiderzee. New inlets formed, allowing the sea to enter into the peat reclamation areas. Tidal action scoured drainage channels. The inland lakes soon became connected to the Zuiderzee.

This was a period of constant battle with the sea. Many dikes were built to protect the towns in the region. Eventually some of these dikes were connected to form regional protection zones. Sea walls were constructed along the Zuiderzee coast. **Figure 1-10** shows the configuration of this region by the year 1250 along with a comparison to the same area in 800. Note the expansion of Lake Almere (also seen in **Figure 1-6**) into the Zuiderzee. Also note the formation of the IJ (pronounced "eye") as a southern appendage to the Zuiderzee. At this same time, another inland lake was growing in size due to land subsidence and peat cutting, the Haarlemmermeer.

Developments in Friesland and Groningen

Much of the area east of the Vlie (**Figure 1-6**) is now part of the provinces of Friesland and Groningen. Peat reclamations also occurred here. As experienced elsewhere, peat reclamation resulted in subsidence of the land. The salt marshes protected the northern areas from direct communication with the North Sea. The only significant loss of land occurred in the region of the Lauwers and Eems inlets as well as the lakes that formed in the southwest part of Friesland.

Figure 1-11: Arrangement of the first river polders.
Redrawn from van de Ven 2004, Ch. 2, Fig. 32.

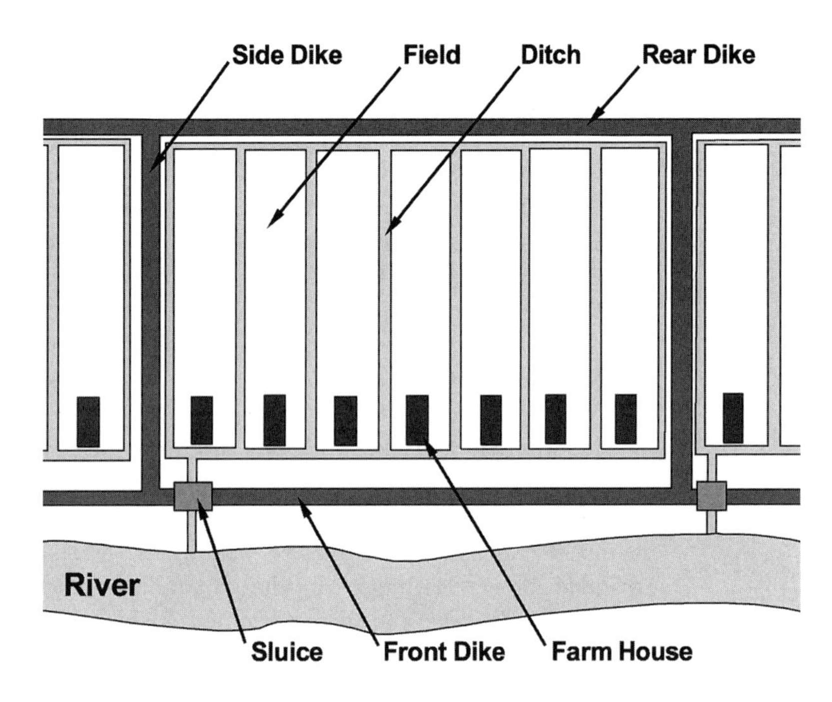

Prior to the Middle Ages the salt marsh region along the northern coast had been populated by those living on the dwelling mounds. These communities continued to grow during the Middle Ages. During this time period dikes were constructed between the mounds. These dikes kept the flooding away, thereby making the dwelling mounds redundant, and by the thirteenth century the northern communities were no longer building or raising any mounds. A more complete discussion of dwelling mounds is the topic of Chapter 2.

Developments in the Land along the Rivers

Peat reclamations along the large rivers followed a different pattern. Low dikes were built along the river. Drainage ditches were dug along the parcels perpendicular to the river and intersecting with a ditch that paralleled the dike. This ditch emptied into the river by way of a discharge sluice. A group of farms was then protected from flooding by two side dikes and a rear dike as shown in **Figure 1-11**. These river reclamations, some constructed as early as 1250, were some of the earliest true polders. A polder, as described earlier, is an area surrounded by dikes within which the groundwater level is artificially controlled.

In the twelfth century dikes were constructed along the lower reaches of the rivers. The low dikes were raised and connected to form continuous dikes. The Lek and the Hollandse IJssel had complete dikes by 1150. The Oude Maas, Merwede, and Maas had dikes by 1250.

The region began to depend more and more on complex drainage systems. The beginning of the eleventh century saw the construction of the first dam across a river. In general, these dams were used to control drainage on a regional scale. A number of dams were built at the upstream end of branching rivers. By cutting off flows from the parent stream, the branch was now able to provide better drainage for the surrounding area. This was done in 1122 on the Kromme Rijn.

At this time many of the peat reclamation areas along the larger rivers were protected by a continuous river dike system. Some of these areas were also drained by smaller natural streams and rivers. The outlets of these smaller rivers created a weak spot in the dike system of the larger river. As a result, smaller rivers needed to be dammed at their confluence with the larger river. These dams were equipped with outlet sluices to allow the smaller river to continue to flow. Many Dutch towns have taken their names from these dams. The dam on the Rotte River (built in 1240) is now the site of the city of Rotterdam, while the dam on the Amstel River (built between 1265 and 1275) is now the site of the city of Amsterdam.

In addition to damming rivers, it was necessary to build regional drainage canals to accommodate the drainage needs of the drained peat lands. In the twelfth century the old outlet of the Rhine River began to fill in. This was partly associated with the formation of new dunes along the coast. The main flow of the Rhine shifted to the Lek and Waal. This meant that the Old Rhine was no longer a viable route for peat-land drainage.

Developments in the Southwest Delta Region

The region at the mouths of the Maas and Schelde Rivers changed drastically during the Middle Ages. During sea transgressions the peat in this region became soaked with salt water and then overlain by marine clay. The salt-soaked peat was a valuable natural resource for the region. The peat would be cut away and salt could be extracted from the peat. By the sixteenth century it was estimated that about 500 square kilometers (200 square miles) had been dug away. The tidal salt marshes in this area were also valuable for grazing, and the raised creek beds provided farming.

The Southwest delta region changed drastically, especially in the eleventh and twelfth centuries. Increasing sea activity coupled with land subsidence and peat removal resulted in the broadening of existing sea arms and rivers as well as the development of new sea arms. Many of the tidal salt marshes that had developed between 300 CE and 800 CE were destroyed.

Yet during this period there were also some increases in land mass, primarily due to natural processes. A great amount of sand moved with the tidal flows along this part of the coast. Some of this sand was deposited, forming both barriers (in the middle of a sea arm) and accretions (along the bank). These barriers and accretions were rather stabile. By the end of the twelfth century this region was transformed into a vast network of islands open to the sea.

The earliest attempts at flood protection were the construction of dwelling mounds of 1 to 2 meters (3 to 7 feet) in height. Toward the end of the twelfth century the area's first regional dikes were constructed. Many of the islands were completely encircled by these protective dikes.

Many flood disasters occurred in the later part of the Middle Ages. Because of these floods, the southwest region suffered extensive, permanent loss of land. Much of this loss was associated with poor management practices. Poorly maintained dikes failed more easily. The economic rewards of peat removal were such that a great amount of this valuable commodity was removed from the areas outside the dikes. This reduced the integrity of the dikes and made it more difficult for them to protect the dry land. In one case, peat was harvested to pay for the replacement of a dike that had been lost due to peat removal. **Figure 1-12** shows changes between 1250 and 1600 at the mouth of the Maas River and the Rhine River branches. Note the loss of the Grote Waard after the Saint Elisabeth's Day Flood of 1421.

Technological Innovations

The major technological innovations in the early Middle Ages were the construction of dikes and dams and the development of the discharge sluice. Mechanical drainage driven by human, animal, or wind power did not come until the late Middle Ages. The first windmills—modified versions of mills used to grind corn—appeared early in the fifteenth century.

The dikes of the Middle Ages were generally constructed of locally obtained materials, often clay or clay mixed with peat. Many "sod dikes" were constructed by stacking pieces of sod. Seaweed was also used in some dike construction. Sea dikes were constructed with a gentle slope on the seaward side and a steeper slope on the landside. River dikes were different. The riverside

Figure 1-12: Transformation in the southwest delta region from 1250 to 1600.

Redrawn from van de Ven 2004, Ch. 3, Fig. 2.

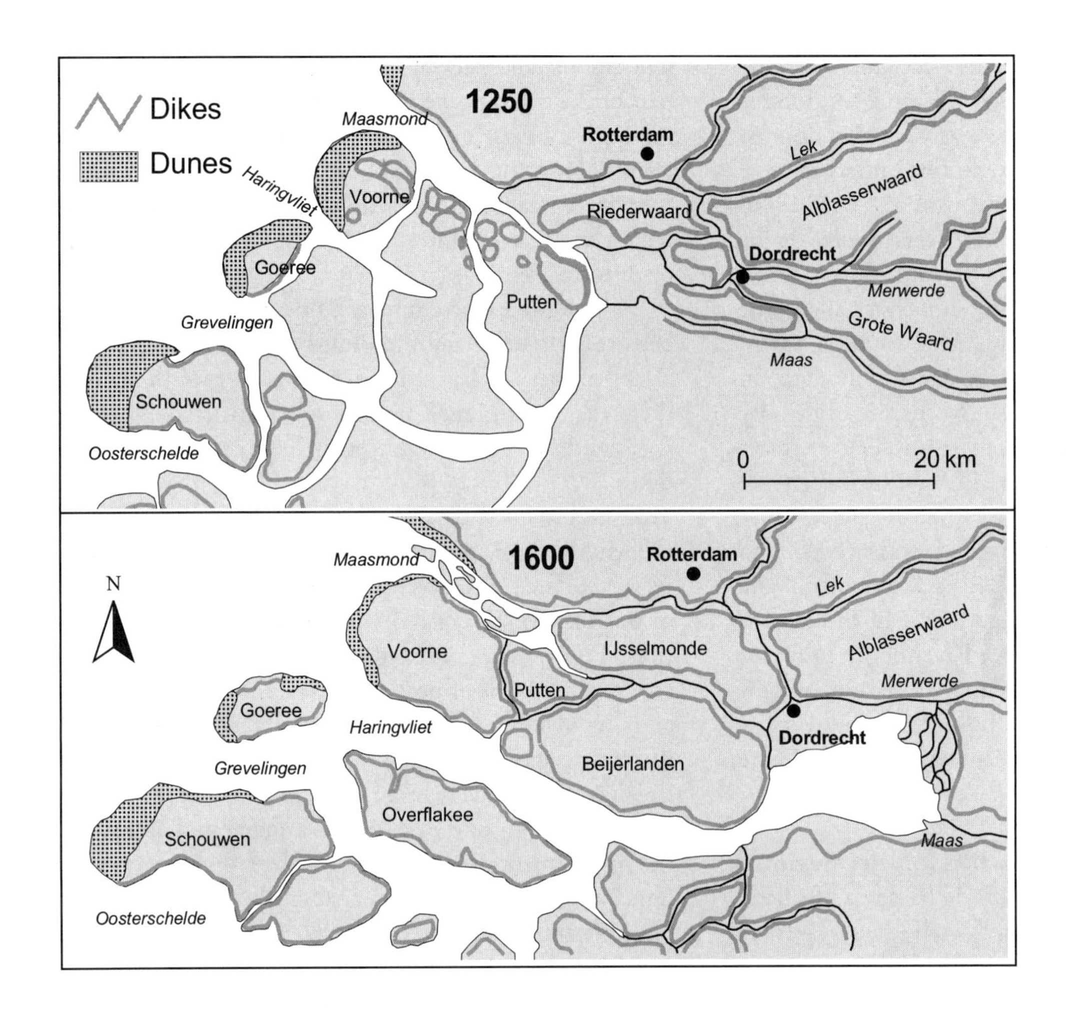

was steep because it had to resist the soaking action of water for a longer period of time. The steeper slope presented a smaller area for water to flow through, making it less vulnerable to damage by seepage.

The tools for dike construction included sod cutters, sod lifters, and wheelbarrows. The borrow materials used to construct the dike were taken from a site as close as possible, as it was extremely difficult to move large volumes of soil by hand. The wheelbarrows were pushed along wooden planks placed on the soft soil. Dike builders developed a "fire brigade" system for the delivery of fill materials. Individuals would push a full wheelbarrow along a plank for a short distance. They would then transfer the full wheelbarrow to the next person in line and return with an empty wheelbarrow. If a greater delivery rate was needed (in case of impending flood season), men were added and the delivery distance was shortened for each person.

The discharge sluice was designed to allow water to drain from within a diked area. It was a simple gate or door that opened to let the water drain when the level outside the dike was low but closed when the outside water rose. This allowed low-lying areas near the coast to drain to the low tide level. Early discharge sluices were simply wooden conduits constructed in the dike body with a gate (vertical hinge) or valve (horizontal hinge).

Figure 1-13: Sluice gate details. The upper two images show cross sections of sluices without and with sheet piling for side and underflow protection. The lower image shows a longitudinal section with valve.

Reprinted, by permission, from van de Ven 2004.

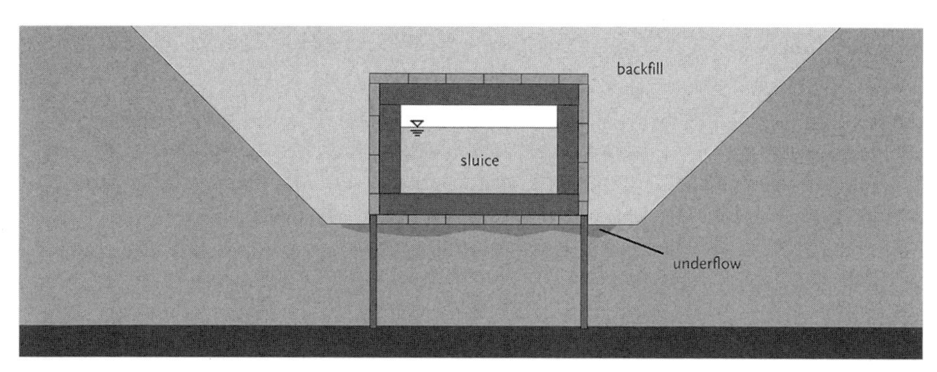

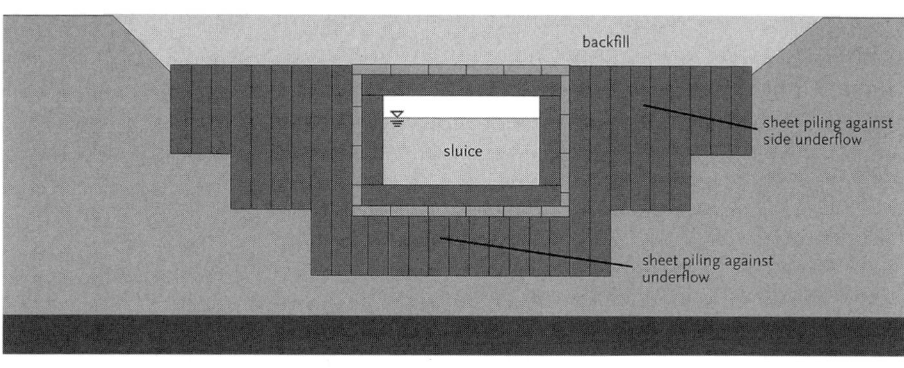

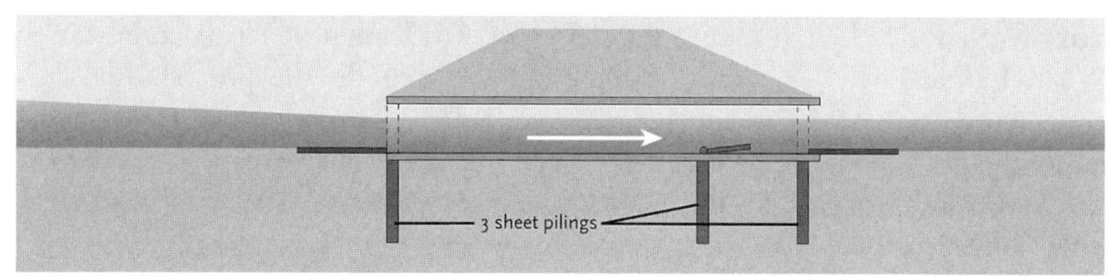

Water inside the dike would open the gate, and water outside the dike would close the gate. **Figure 1-13** shows one of these sluices. The sluice was constructed in a dry construction pit, and the discharge channel was diverted through it after construction. These sluices suffered many failures. Some were damaged when their gates slammed closed. Others were undermined by seepage. Often primitive sheet piles were installed to solve this problem (see **Figure 1-13**). The construction of the sluice was often hampered by upward seepage in the bottom of the construction pit.

Perhaps the most significant technological development in land reclamation during the Middle Ages occurred later in this era. By the early part of the fifteenth century some of the peat soils had subsided to a level where it was no longer possible to keep land dry by gravity-based technologies alone. It was then that the Dutch began to use the windmill for pumping water. Windmills had already been used for grinding corn, so windmill technology had already reached a high level prior to its use in water drainage. The first record of a windmill constructed for water drainage was near Alkmaar in 1408. The first windmills used scoop wheels to lift the water. Initially, few windmills were built due to the high cost of construction. By the seventeenth century the country had become much more prosperous, and windmills were constructed at a rapid pace. Chapter 3, which deals with seventeenth century reclamations, covers the development of windmills in more detail.

National Independence and the Golden Age

The sixteenth and seventeenth centuries were important in Dutch history. In this period, the Dutch provinces gained their independence from Spain and attained great wealth and prosperity. The seventeenth century, in particular, became known as the "Golden Age" when wealth was attained, in large measure, by international trade from the shipping industry. This was also a period of great advancements in land reclamation when the Dutch people developed new technologies to keep their feet dry.

Independence from Spain

The first half of the sixteenth century saw the beginnings of the Protestant reformation with the teachings of Martin Luther and John Calvin. Calvin's ideals were attractive to a world characterized by urban development, trade, and manufacturing. Where Luther advocated a return to primitive simplicity, Calvin's ideals of thrift, industry, sobriety, and responsibility promoted economic development. Calvin saw the state as subordinate to the church. He also believed that lower levels of government should oppose any ruling tyrant. The teachings of Calvin gained popularity in the Dutch provinces, especially in the second half of the sixteenth century, despite the Inquisition against Protestants initiated by Charles V.

In 1555 Philip II took over power from his father, Charles V. Ruling from Spain, this absentee monarch demanded loyalty to the crown and to the Roman Catholic Church. Discontent among many members of Dutch society grew. The lesser nobles did not like being under the rule of a foreign king and the restrictions that were being placed on their independence. The merchants believed that they would be more profitable without Spanish rule. The Calvinists wanted to worship in freedom and spread their beliefs.

In 1566 a group of lesser Dutch nobles signed a pact called the "Compromis," in which they stated their opposition to policies of Spanish rule. At this same time, Calvinist rebels started riots in an "iconoclastic fury" against the Catholic Church. The first armed conflict against Spanish rule took place in the spring of 1567. Later that year Philip II sent the Duke of Alva, Fernando Alvarez de Toledo, to the Dutch provinces with an army to defend the Spanish crown and the Catholic faith.

One of the prominent leaders of the time was William of Nassau (or William of Orange), a wealthy landowner and also the ruler of Orange, an independent principality in the south of France. In 1559 he was appointed by Philip II to be the stadholder of the provinces of Holland, Zeeland, and Utrecht. He represented these provinces in the Council of State. Despite being Catholic and an appointee of the state, he voiced his opposition to the absolute power of the monarch. In 1567, as the revolt was heating up, the nobles were asked to swear an oath of loyalty to Philip II. William's refusal eventually made him the leader of the opposition. He and other Calvinist rebels were forced to flee to Germany. The Duke of Alva's reign of terror lasted for six years—from 1567 to 1573. In the process thousands were executed.

After several unsuccessful attempts, forces loyal to William of Orange were able to take control of some Dutch coastal towns in 1572. Water was first used as a military defensive tool in 1573. When the Spanish tried to take the city of Alkmaar, the residents flooded the surrounding fields by cutting dikes, forcing the Spanish forces to retreat.

In 1579, delegates from one of the southern provinces (now part of Belgium) declared their loyalty to the king. In response, several northern provinces decided to establish a closer union to continue their fight against Spain. The agreement that was signed, called the Union of Utrecht, was the foundation of a new republic, called the Seven United Netherlands. It consisted of seven provinces (Holland, Zeeland, Utrecht, Friesland, Groningen, Overijssel, and Gelderland) joined into a confederation where joint decisions were limited to issues of defense and foreign affairs. These decisions were made by the States-General. Most of the power lay in the provincial councils composed of nobility and representatives from the cities. The provincial executive was the stadholder, who was now appointed by the States-General instead of the king. In 1581 the States-General formally declared its independence from Spain.

In 1584 the leader of the revolt, William of Orange, was assassinated. One year later his son, Maurits, became the stadholder of the two most powerful provinces, Holland and Zeeland. He was able to organize the rebellion into a coherent, successful revolt. Under Maurits's leadership, progress was made in the war against the Spanish. Maurits died in 1625.

Under the leadership of Fredrick Henry, Maurits's brother, the Dutch made further gains against the Spanish. Finally, in 1648 a peace treaty was signed that officially ended 80 years of conflict with Spain and established the Republic of the Seven United Netherlands as a free and sovereign country.

The Golden Age

The prosperity of the Golden Age can be traced back to the 1580s. About that time the cities of Amsterdam and Middelburg (as well as the provinces of Holland and Zeeland) grew to become centers of world trade. Their dominance in trade was strengthened by the Spanish capture of Antwerp in 1585 and the resulting exclusion of merchant shipping into that city's port.

Dutch merchant ships traded salt, wine, spices, herring, and textiles with the Baltic region. They shipped grain to Spain and brought goods from Persia to Western Europe via trading partners in Italy. Yet the most important trading activity was with the Far East. In 1602 the Dutch established the Dutch East Indies Company or (in Dutch) the *Verenigde Oostindische Compagnie* (VOC). Their trading routes passed the Cape of Good Hope, where they set up a staging post for en-route supplies (now South Africa). They traded extensively with Java, where they established the town of Batavia (now Djakarta, Indonesia). The Dutch also traded tobacco and sugar in Recife, Brazil. In 1609 Henry Hudson, an English explorer, was hired by the Dutch East Indies Company to find a new route to the Far East. Instead, he established the village of New Amsterdam (now New York) in order to trade with Native Americans. The Dutch were active in the slave trade as well, bringing African slaves to North America.

Most of the Golden Age wealth was concentrated in two provinces, Holland (including the city of Amsterdam) and Zeeland. At the height of the Golden Age, around 1640, the Dutch had approximately 2,500 vessels actively involved in trading activities (plus at least 2,000 additional fishing vessels). A period of slow decline occurred toward the end of the seventeenth century. One reason was that England, the main competition for international sea trade, passed a series of laws that excluded Dutch ships from English harbors.

The Golden Age saw the end of the Dutch revolt (discussed earlier) as well as increasing conflicts with France and England. France, under the leadership of Louis XIV, was trying to establish itself as the leading continental power. England was trying to take away the Dutch trade monopoly. In 1672, as the Netherlands became involved in wars with both England and France, William III (great grandson of William of Orange) was appointed to the position of stadholder (of five provinces) to lead the country in this time of conflict. Later, in 1688, William III became king of England through marriage to Mary Stewart, daughter of King James II. As king of England, he strengthened the English navy, creating further loss in Dutch dominance at sea.

Land Reclamation

The beginning of the seventeenth century marked an important transition in the area of drainage, flood protection, and land reclamation. The Dutch lowlands had been subsiding since the year 1000, primarily as a result of agricultural drainage activities. In addition, the level of the North Sea was steadily rising. Up to about 1500 the force of gravity was sufficient for most land drainage. Areas could be kept dry using dikes and outlet sluices. Around 1500 many areas had subsided to the point where gravity drainage was no longer possible and mechanical devices were required. **Figure 1-14** shows this transition.

The Golden Age was a very significant time for the development of land reclamation technologies. Early in the seventeenth century a number of lakes that had formed north of Amsterdam were drained and reclaimed to become valuable farmland. Several things came together at the same time to allow this to happen. The lakes north of Amsterdam were growing larger with each

Figure 1-14: Sea level rise and land surface fall since 1000 CE.

Reprinted, by permission, from van de Ven 2004.

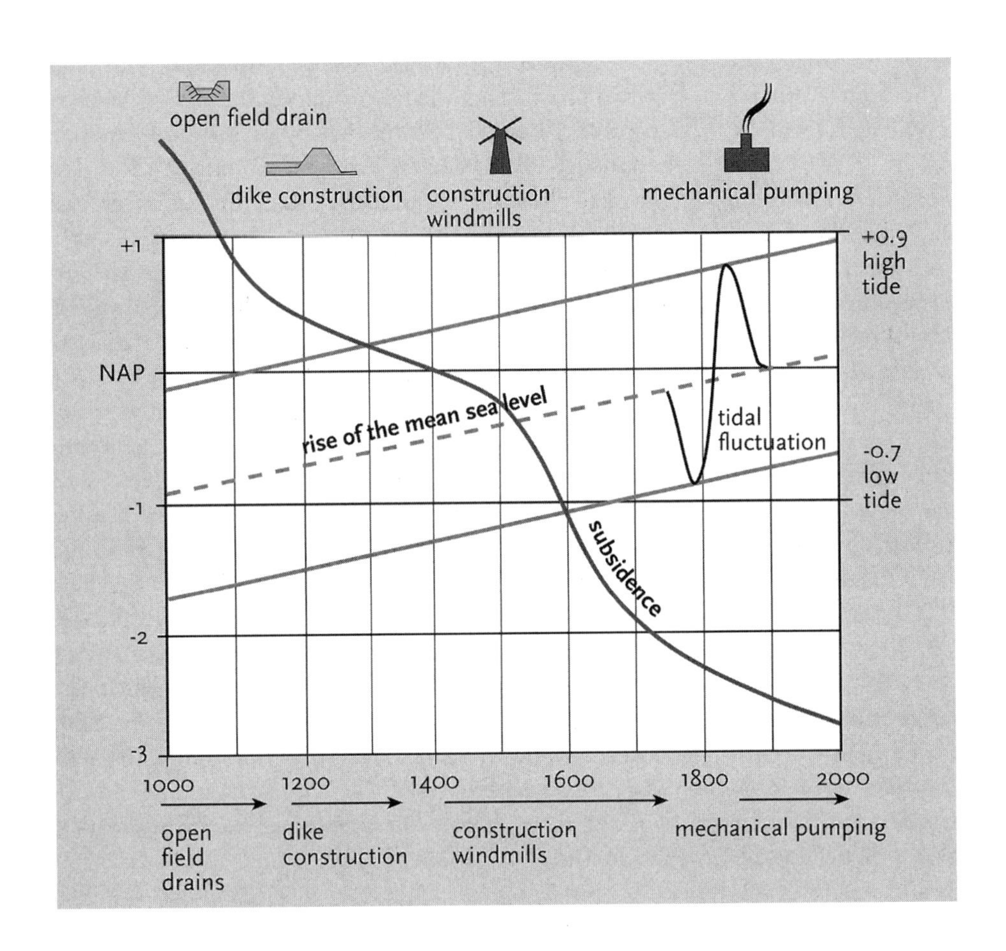

passing storm. It was feared that much of this region would eventually be washed away. Windmill pumping technologies had developed to the state where they could actually drain an entire lake if enough windmills were used simultaneously. Consequently, any large lake drainage project would require significant financial resources. Because of Golden Age wealth these resources were available. Drainage of the lakes in the vicinity of the rapidly growing cities was also stimulated by the growing need for agricultural land in order to produce food.

The process of draining an entire lake started with the construction of a pair of dikes surrounding the lake. The space between the dikes would become the canal from which the drained water would eventually flow to the sea. A large network of windmills was constructed around the perimeter of the lake. Because of the depth of such a lake, the mills were often arranged in series, each successive mill lifting the water higher. As the lake drained, a system of drainage ditches was put in place, allowing efficient drainage to the windmills. This will be covered in more detail in Chapter 3.

The prosperity of the Golden Age also brought a change to the activities of the water boards. Farmers who wanted to focus their attention on agricultural production were no longer willing to actively maintain the dikes and ditches. Instead, the water boards began to hire professional labor for this purpose. This resulted in a new industry led by trained hydraulic engineers.

Eighteenth and Nineteenth Centuries

The eighteenth century was a period of economic decline, warfare, and political upheaval for the Dutch republic. The nineteenth century brought new developments primarily in the areas of industry and colonialization. Technologically, the most significant development was the use of steam power for drainage.

Historical Developments

The economic decline in the Netherlands during the eighteenth and nineteenth centuries was due to a number of factors. The Dutch had lost their hold on their trade and finance monopolies of the seventeenth century, and they continued to become involved in frequent conflicts with neighboring countries. In this period there were several major changes in governance.

The political system in place through the end of the eighteenth century was initially established in 1579 with the Union of Utrecht. The main political struggle was between members of the House of Orange, who held the position of stadholder in the principal provinces, and the administrators of the Holland towns. There were several periods, totaling 70 years, when, because of the power of the towns, no stadholder was appointed. By the end of the eighteenth century, many in the republic believed that the existing system was no longer working. The patriot movement, encouraged by the American Revolution, promoted a change to a more democratic government. They blamed the stadholdership for all of the country's troubles. Change came at the end of the century.

In 1795 the French successfully invaded the country. In an effort to please the French, the States-General repealed the stadholdership and formed the Batavian Republic (named after the Roman-era inhabitants). But, after ten years, little progress was made at building a constitutional republic following the ideals of the Patriots. As a result, Napoleon replaced the Batavian Republic with the Kingdom of the Netherlands, ruled by his brother Louis Napoleon Bonaparte. When Louis renounced the throne in 1810, Napoleon annexed the Netherlands.

French rule united the country. It created uniform governance throughout the Netherlands with many government activities (taxation, education, etc.) becoming centralized. But, the French also restricted trade activities, ruining that part of the economy. When Napoleon was defeated in 1813, a new government was formed by proponents of a monarchy. They selected another member of the House of Orange to run the country—this time as King.

King William I actively tried to return the country to its former prosperity. He had a long list of accomplishments. He established colonial rule in the East Indies, including sending Dutch manufactured goods east. He encouraged industry and began an effort to expand the road network in the country. He constructed the Noord Hollands Kanaal, a canal giving larger ships a more dependable route from Amsterdam to the North Sea. He built the first Dutch railroads and established the Academy for Civil Engineering in Delft (now Technical University Delft).

In 1815, under the kingship of William I, the seventeen provinces were reunited as the Kingdom of the Netherlands. (The southern provinces split in 1579.) This reunification was not successful. Cultural, language, economic, and political differences caused the union to split again in 1839. The southern provinces became the countries of Belgium and Luxemburg.

When King William II took control from his father in 1840, many in the Netherlands again began to oppose rule by a monarch. In response to this opposition, a new constitution was written and adopted in 1848, peacefully turning the Netherlands into a democratic, constitutional monarchy. The new constitution provided for the direct election of a Lower House. Members of the Upper House were elected by the provincial councils. It opened all meetings of representative assemblies to the public. Furthermore, under this constitution, ministers were to be answerable for the actions of government instead of the monarch. The new constitution also introduced freedom of education, assembly, expression, and religion. With this new constitution the king's powers were greatly limited. Johan Rudolf Thorbecke was the architect of the new constitution and became the first prime minister under the new government.

William III became king in 1849 and oversaw a period of significant industrial development. In 1863 a system of university preparatory schools were established. In the 1870s and 1880s some of the first Dutch political parties were formed. King William III died in 1890. His daughter and heir, Wilhelmina, was only ten years old at the time and too young to succeed him. Her mother, Emma, became regent until 1898 when, at the age of eighteen, Wilhelmina became the first queen of the Netherlands.

Technological Developments

One of the more notable engineers of the eighteenth century was Nicolaas Cruquius (1678–1754). Cruquius was one of the first to bring scientific principles into the practice of hydraulic engineering. He began his career in 1698 as a surveyor. In 1717, at the age of 39, he enrolled at the University of Leiden. Cruquius learned that much could be gained by careful detailed study. In 1728 he was asked to find a way to prevent erosion along the coast of the island of Goeree in the southwest delta region (see map in **Figure 1-12**). Instead of proposing to build defensive works, he suggested that the currents in the region be investigated. Four years later, he proposed the construction of a dam that would change the direction of the ocean currents, thereby reducing the erosive forces along the Goeree coast. Cruquius also convinced the Rijnland water board to begin collecting meteorological data. Beginning in 1735, measurements were made of air pressure, wind force and direction, evaporation, precipitation, and icing. These are now some of the oldest weather observations in Europe.

During the French rule, there was a shift toward more centralized control. This was also true in the area of water management. In 1798 a national administration, called Rijkswaterstaat, was established to provide technical hydraulic engineering support. After the Netherlands became a kingdom, under the reign of the House of Orange, the new government remained a strong centralized state, establishing a permanent role for centralized water management.

The most significant technological development of the eighteenth and nineteenth centuries was the use of steam power for drainage. Steam power had developed in England in the eighteenth century for the purpose of draining mines. As early as 1710, a practical steam engine was developed for draining mines. It took many years for this technology to be used in the Netherlands for drainage. There were several reasons for the delay. First, the Dutch successfully used windmills to keep their feet dry for several centuries. Furthermore, wind energy was free while coal had to be imported. Second, there were technological challenges. The early steam engines were designed for high lift and low volume conditions. Drainage required low lift and high volume pumps. In addition, there were problems associated with pumping under varying lift conditions. The first successful steam application in the Netherlands was in the city of Rotterdam in 1776.

The use of steam technology progressed very slowly in the Netherlands until a decision was made in the middle of the nineteenth century to drain the ever-growing Haarlemmermeer. By 1848 this lake had grown to 180 square kilometers (70 square miles) with an average depth of 4.5 meters (15 feet). Eight-hundred million cubic meters (28.3 billion cubic feet) of water had to be drained. This would have been a monumental task for windmills alone. After much discussion, a decision was made to drain this lake using three steam-driven pumps. These pumps were the largest steam-driven pumps ever built. The Haarlemmermeer was finally drained in 1852. Today the country's major international airport—Schiphol, named after shallows in the lake—lies in this drained lake. The development of steam pumping and draining of the Haarlemmermeer is covered in Chapter 4.

Twentieth Century

The twentieth century was a period of significant economic growth, two world wars, several flood disasters, and the two largest land reclamation and flood protection projects ever undertaken. The first was the closing and partial reclamation of the Zuiderzee, and the second was the closing of the estuaries in the southwest delta.

Historical Developments

The turn of the twentieth century was generally a prosperous time for the Netherlands. It was a time when European power was strongly divided between the German empire and the Britain-France-Russia bloc. The Netherlands managed to remain neutral through World War I. The Germans were given permission to ship goods through the country but were not allowed access to the North Sea. When the Germans eventually invaded Belgium (another neutral country) to the south, economic crises arose in the Netherlands.

Beginning in 1929, the Netherlands, along with most of the rest of the world, suffered through the Great Depression. By 1935, unemployment was running at 40 percent.

The Netherlands and its people also suffered greatly through World War II. The Germans invaded on May 10, 1940. German troops attempted to capture the members of the government and the royal family. With the help of the Dutch military, Queen Wilhelmina and her cabinet were able to escape to England where they established a government in exile. On May 14 the center of Rotterdam was leveled by a German bombing attack. When the Germans threatened to do the same to Utrecht, the Dutch army surrendered, and Germany occupied the Netherlands.

On June 6, 1944 the Allies invaded Normandy, giving new hope that the occupation would soon be over. By the fall, the Allied forces reached Belgium, and not long afterwards the southern Dutch provinces were liberated. Several months later, on May 5, 1945, almost a year after the D-Day invasion in France, the rest of the Netherlands was liberated. The cost of the war was great. About a third of all industry was destroyed, along with 60 percent of the transport system. Much of the housing was not usable. The recovery was aided by economic assistance provided under the Marshall Plan in 1948. The postwar period also saw the institution of new political parties. In 1948 Queen Juliana took over the role of queen from her mother. In 1980, Juliana's daughter, Beatrix, became queen.

By 1960 the Netherlands was a fully industrialized country. Agricultural production increased greatly, and Rotterdam now achieved the status of being the largest port in the world. Large, multinational companies such as Philips, Shell, and Unilever became major employers. Added to this picture of growth and success was the discovery of one of the largest natural gas reserves in the world in the Province of Groningen.

One reason for the more recent economic success was the consensus-based economy often referred to as the "polder model." Life in the early Dutch polders could not be sustained without the cooperation of all parties involved. This required constant consultation between the public, the landowners, and the water boards. It was necessary that each party at least understand the point of view of the others. As a result, consensus seeking is part of the Dutch character, and their economic system gives evidence of this trait. It features close cooperation between trade unions, government, and the private sector. This approach produced an economic growth in the 1990s greater than the average for countries of the European Union.

Flood Protection and Land Reclamation

The flood protection and land reclamation projects of the twentieth century attained a new level of scale and technical merit. Two projects deserve attention.

The reclamation of the Zuiderzee was the largest single land reclamation project ever completed. After much discussion at the end of the nineteenth century, after flooding threatened Amsterdam in 1916, and after food shortages during World War I, a decision was made to cut off the Zuiderzee from the North Sea and reclaim a large portion of the sea bottom. First, the sea arm was dammed, forming a freshwater lake—now called the IJsselmeer (Lake IJssel). Next, dikes were constructed within this lake and water was pumped out from within the diked area, creating 1,650 square kilometers (640 square miles) of new dry sea-bottom lands. After years of preparation, the land was made useful for agriculture. In addition to agricultural lands, new towns and cities were designed and built "from scratch." The largest of these new cities, Almere, is still under construction today. The reclamation of the Zuiderzee is covered in Chapter 5.

In 1953 the Dutch experienced one of the worst flood disasters in their long history of working hard to keep their feet dry. The southwest delta region was hit by a storm surge, producing flood levels higher than ever recorded. The damages included 800 kilometers (500

Figure 1-15: Gains and losses of land through natural causes and reclamation since 800 CE.

Redrawn from SWAVN 1984, Fig. 6, and IDG 1994, p. 9.

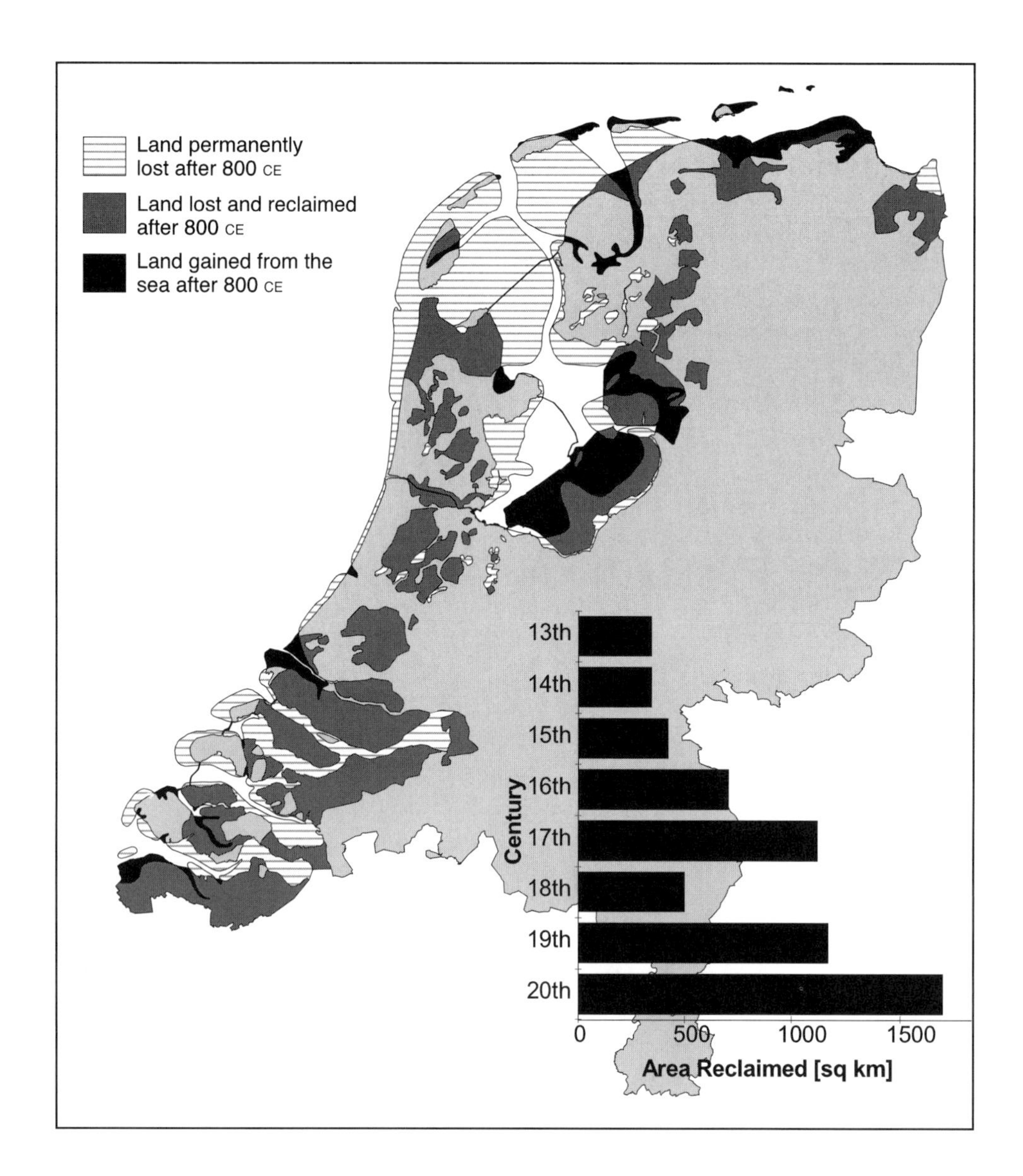

miles) of dike severely damaged, 2,000 square kilometers (770 square miles) of land inundated, 3,000 houses destroyed, and 43,000 houses damaged. There were over 1,800 human lives lost. In response to this disaster, the country initiated the single largest flood protection project ever constructed in the Netherlands. This project, referred to as the Delta Plan, involved damming many tidal estuaries and constructing several storm barriers. The details are provided in Chapter 8.

Figure 1-15 summarizes the coastline changes and land reclamation activities since the year 800. Most of the reclaimed land was dry land in 800 and then was lost due to mismanagement, storms, rising sea levels, or a combination of these. Many areas were lost and never reclaimed.

2

Dwelling Mounds: An Escape from the Floods

The earliest form of protection from flooding involved the use of raised dwelling mounds. This technique was not very complex, and it really did not require any real innovation. Dwelling mounds developed quite naturally in the process of people simply trying to keep their feet dry. It is important to note that this early form of civil engineering had a significant impact on the society and prosperity of the northern coastal regions. This chapter covers the development of dwelling mounds and the effect they had on the people of the region.

Dwelling mounds can be found throughout the region of Holocene deposits along the North Sea—all the way from the Netherlands to Denmark. As already mentioned in Chapter 1, they go by various names. In the province of Friesland they are referred to as terps. In the province of Groningen they are called wierde. The focus here is the coastal region in the present Friesland and Groningen provinces. In particular, we focus on the region along what is now known as the Waddenzee (or Wadden Sea).

Description of the Wadden Region around 500 BCE

Since the last ice age the sea level has risen approximately 100 meters (330 feet). During this process the straits of Dover were flooded, separating what is now England from the rest of the European continent and significantly changing the currents in the North Sea. These currents began to run north parallel to the Dutch coast, resulting in the growth of barrier islands and sand banks. The barrier islands were able to grow in this region partly because of limited tidal fluctuations. The inland groundwater levels also rose with the sea level. As a result, a broad belt of marsh and bog spread over the gradually sloped Pleistocene surface. The marsh grew so thick that it became isolated from the river water. One plant that can survive in this environment is sphagnum moss, which eventually forms a thick peat layer. At one point the peatlands stretched from the coastal dunes to the higher Pleistocene ground in southeast Friesland and Drenthe. Between the sandy barrier islands and this inland bog was a large tidal lagoon of mud flats and shallows—now called the Waddenzee (reference **Figure 1-5**). Sediment was transported through the tidal inlets. Sand was deposited along the banks, and finer clay particles were transported further inland. The result was the formation of an elevated salt marsh fringe along the lagoon and inlets. This salt marsh fringe provided protection for the inland bog. In many locations the subsurface is composed of alternating layers of peat and marine clay, created by variations in peat growth and water level rise.

Tidal movement, extreme weather conditions, continuous erosion, sedimentation, and peat growth created a very dynamic environment. This was not an easy place for man to settle.

Early Settlement in the North

The earliest permanent settlements along the Dutch Waddenzee coast date from around 600 BCE (Vollmer et al. 2001). Prior to this time, some residents of the higher Drenthe plateau located 25 to 50 kilometers (15 to 31 miles) south of the coast used this area in the summer but then retreated to higher ground in the winter. The first settlements were along the salt marsh fringe just above high tide. Within a couple of generations the settlers realized that with a bit of added elevation they could protect themselves from flooding caused by storms. By piling human and animal wastes along with sod they were able to construct small mounds that would provide this protection. Initially this mounding of debris came about simply as a part of human habitation. It was not originally intended to be a method of flood protection. But as the sea level began to rise further, it was clear that the higher mounds had the greatest security. In time any new farmstead construction began with the construction of a dwelling mound.

During times of increased marine activity, some of the mounds were raised. Others were simply abandoned. Raising a mound was done with manure and grass sod. These mounds grew with time, and in several cases neighboring mounds coalesced into one mound. This was the beginning of a village mound. During the winter the village mound provided the needed protection to store the winter stocks. During the summer the farmers cultivated crops such as barley, millet, flax, or emmer in the land surrounding the mound. These crops were selected because this cultivated land was frequently flooded with salt water. Drainage ditches were dug in the cultivated areas with drainage patterns extending radially outward from the mound to the tidal creeks.

Dwelling mounds often had a similar arrangement. At the center there was a small pond or *dobbe* dug to collect fresh rainwater. The farmhouses were then oriented with the living part around the (higher) center of the mound and the stable farther down the slope, allowing good access to the fields surrounding the mound. A ring road was often located at the base of the mound, connecting the farms together.

Caius Plinius Secundus, otherwise known as Pliny the Elder, served in the Roman army in the first century CE. He visited the Dutch coastal region in 77 CE and wrote about his experiences in his 37 volume work *Naturalis Historia*. He described life along the Waddenzee coast as follows (Meijer 1996):

> Day and night save for two lulls, the great expanse of the ocean with its immense waves hurls itself against the land, leading the spectator to this eternal struggle between the elements to wonder whether the ground belongs to land or sea.
>
> There dwell an unfortunate people on hills, or rather on rises that they have erected with their own hands to a height which experience has shown them corresponds to the highest tide. Their huts stand on these rises. They resemble seafarers when the sea submerges the surrounding land and castaways when the waters recede and they pursue the fish which try to flee with the waters.
>
> They cannot keep livestock, and cannot therefore feed on milk, like their neighbors. Nor can they catch wild animals, for the sea has swept away all concealment as far as the eye can see. They make fishing nets from ropes made of woven reeds and rushes. They cut sods of earth with their bare hands which they then dry—in the wind more often than the sun—and burn them to cook their food and to warm their limbs numbed by the north wind. Their only drinking water is rainwater collected in a pit by the entrance to their dwelling.

After being impressed by the way the dwelling mound settlers dealt with their harsh natural environment, Pliny lost his own life in another "harsh" environment. He died two years later in Pompeii with the eruption of Mount Vesuvius (Spier 1969).

In early Roman times there was extensive trade between this region and the rest of the Roman Empire. The dwelling mound residents traded agricultural products for things such as coins, statues, pottery, and jewelry.

The Roman period brought trouble to this region. Increased sea activity, the introduction of malaria, and political turmoil caused the inhabitants to give up their settlements. By 300 CE some Frisian coastal marshes were abandoned, and the Groningen coastal marshes saw a significant drop in population (Vollmer et al. 2001).

Growth of Dwelling Mounds

Resettlement of the coastal region began around 700 CE. This was, in part, due to the fact that this region came to lie at the crossroads of important trade routes. At this time trading terps developed. These were oblong shaped dwelling mounds located along the tidal creeks. Instead of being populated by farmers, they were populated by merchants and craftsmen.

The trading activities led to a period of wealth, which continued into the twelfth century. The number of Romanesque churches and monasteries gives evidence to this.

Dwelling mounds varied in size. A mound large enough for a single farmstead was as small as 40 meters (131 feet) in diameter. Neighboring mounds often coalesced into larger mounds, rising to heights above 6 meters (19 feet). The tallest dwelling mound is Hogebeintum in the province of Friesland. It reached a height of over 8 meters (26 feet) above sea level and nearly 10 hectares (25 acres) in size.

Dike construction started early. The primary purpose was to protect the fields surrounding the dwelling mounds. There is evidence of dikes constructed in the pre-Roman Iron Age. In the tenth century dike construction increased. The process began with a few neighbors building ring dikes around their fields. Later individual villages or groups of villages worked together to build dikes around their towns. The entire Westergo and Oostergo regions had complete ring dikes by 1100. Dike construction along the coast and the tidal estuaries started later.

Around the eleventh and twelfth centuries the dikes also started to expand into the surrounding peat lands. Peat reclamation involved constructing ditches to drain the fields and dikes to keep water from flowing in from surrounding lands. This process of reclaiming peat lands is described in more detail in Chapter 3. By the end of the Middle Ages the area of dwelling mounds was completely diked in.

The construction of dikes changed the area significantly. Since the agricultural land was protected from salt water, it was possible to grow different kinds of crops than the salt-resistant crops grown earlier. Dairy production increased because the fields could now produce more grass than before. Farms were no longer restricted to the dwelling mounds.

One of the consequences of extensive dike construction was the loss of open connection with the sea. As a result, the area lost its leading position in trade during the late Middle Ages. This was partially compensated for by the construction of many canals throughout the region to facilitate drainage and trade. An example is the former Delf canal dug around 1200 between the Groningen towns of Delfzijl and Winsum.

Attempts to bring Christianity to the region began in the early Middle Ages by missionaries like Boniface. Eventually the entire area was christianized. As a result, nearly every village built a church. Early churches were wooden structures eventually replaced by brick churches. With the region diked in and more freshwater available, the dobbe were no longer needed. The church claimed the central location once set aside for the dobbe. Most of the existing dwelling mound villages today retain this layout. The elevated, centrally located church has become a recognizable characteristic of the dwelling mound.

Dwelling Mound Construction and Location

Dwelling mounds were constructed and enlarged over a long period of time from the fifth century BCE to around 1200 CE. Archaeological evidence suggests that the construction and enlargement did not happen uniformly over this time. As mentioned earlier, much of the Dutch lowlands were abandoned at the end of the Roman period of occupation. Archaeological and geological evidence suggests that there were four distinct periods of dwelling mound construction. These periods of construction started around 500 BCE, 200 BCE, 700 CE, and 1000 CE (Meijer 1996). The geologic evidence is based on the subsoil on which the mound rests. Over the centuries there were a number of major sea incursions. Each one spread a new layer of marine clay. The layer on which the mound rests then suggests the time period during which construction first began. **Figure 2-1** shows a cross section through the dwelling mound region

Figure 2-1: Cross section showing generations of dwelling mounds.
Redrawn from Boersma 1972, p. 18.

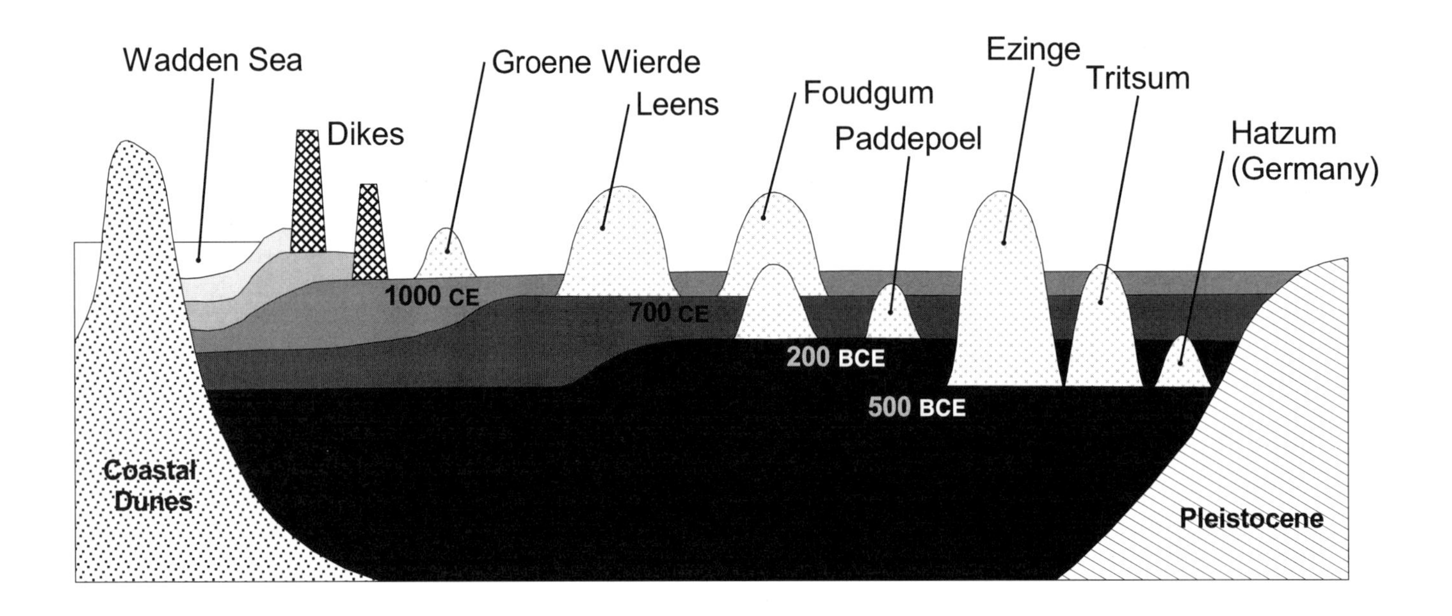

showing this variation in foundation depth. Some of the early mounds that were abandoned are now completely covered. Others that have seen continuous occupation, such as Ezinge, extend through the various layers of marine clay.

The dwelling mounds were most heavily concentrated in the Frisian and Groningen coastal regions. The location of these northern dwelling mounds is shown in **Figure 2-2**. The primary factor that determined the location of dwelling mounds was the location of the salt marsh ridges. These ridges followed the estuaries and rivers.

In the province of Groningen the Fivel River at one time cut a path from the peat areas east of the city of Groningen to the northeast. Around 300 CE the mouth of this river was cut away by marine activity and turned into a broad estuary. The oldest dwelling mounds in this area follow the shore of this estuary. **Figure 2-3** shows the dwelling mounds in this area. The older land is shown in darker shades and newer land in lighter shades. Dwelling mound towns such as Usquert, Rottum, Kantens, Middelstum, and Westerwijtwerd are some of the oldest in this area, and they followed the western side of the Fivel estuary. Loppersum, Leermens, and Godlinze followed the eastern side. By 1250 the Fivel bay was silted in. In 1453 a dike was constructed to complete the process. The bay provided fertile soil, and consequently new dwelling mound villages were constructed following the moving coastline. These included villages such as Stedum, Westeremden, and Walsweer. The Fivel River has since completely disappeared, but it leaves as its legacy a trail of dwelling mound villages.

Friesland's landscape also reveals a pattern of dwelling mound construction. The Middelzee was at one time a tidal estuary at the mouth of the Boorne River, which existed as early as 500 BCE. It was the dividing line between the two Frisian districts of Westergo and Oostergo. This estuary did not begin to silt in until around the year 1000 toward the end of the period of active dwelling mound construction in this region. As a result, no dwelling mounds were ever built in this area. The location of the former Middelzee can be easily identified in the overall map of dwelling mounds in **Figure 2-2**. It is the area with no dwelling mounds directly west of the city of Leeuwarden. (This is shown in more detail in **Figure 7-7**.) The mounds on either side of the former Middelzee are arranged in straight rows following former salt marsh ridges that grew up during various stages.

▾ **Figure 2-2:** Location of dwelling mounds in the northern part of the Netherlands. The dwelling mounds, shown in black, follow the salt marsh ridges and are found primarily in the marine clay areas.

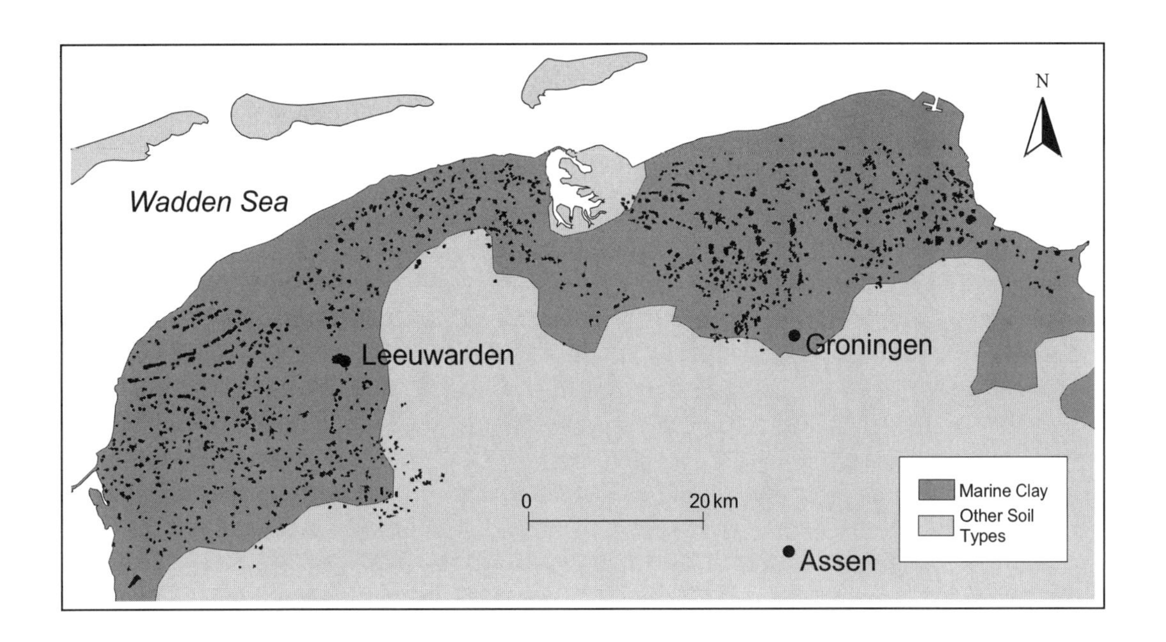

▾▾ **Figure 2-3:** Pattern of dwelling mound growth and later settlements along the old Fivel River. Sea dikes constructed between 1453 and 1944 are also shown.

Redrawn from Elema, Klugkist, and Reinders 1983, Fig. 1.

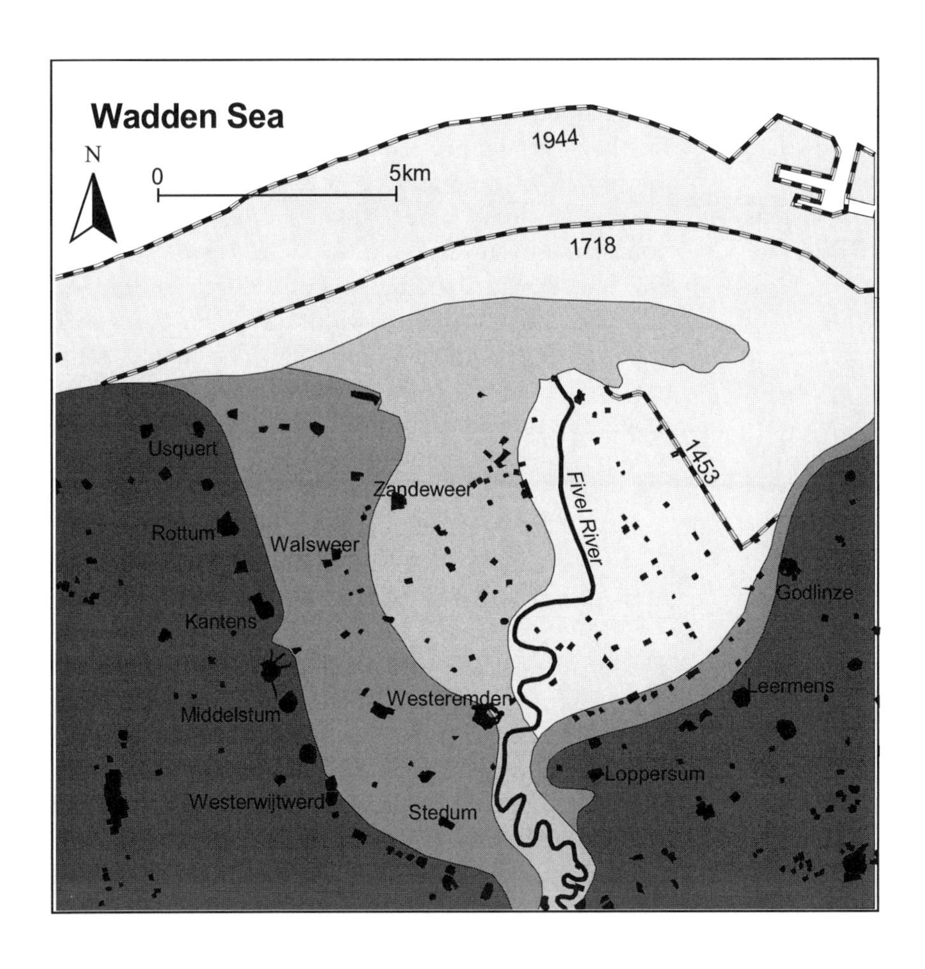

▸ **Figure 2-4:** Excavation of the dwelling mound Eenum in 1925.

From Groningen Institute of Archaeology (GAI); formerly known as BAI.

▾ **Figure 2-5:** Cross section of dwelling mound Hogebeintum.

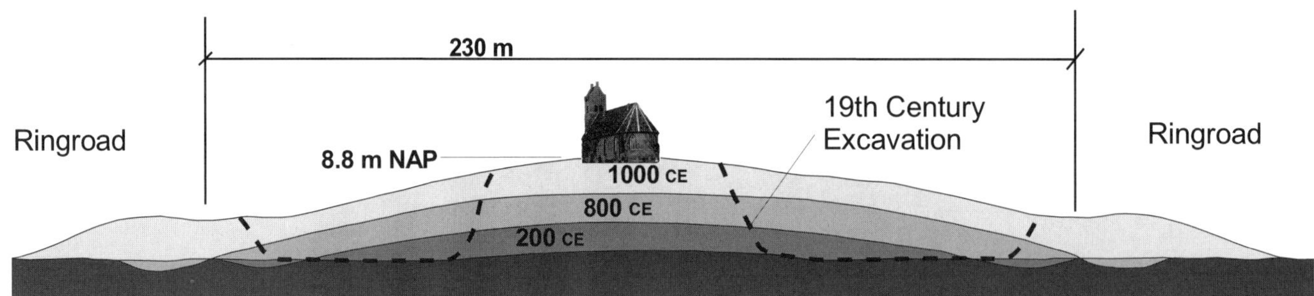

Loss of the Dwelling Mounds

The dwelling mounds have a rich historical heritage. Some of the mounds have seen continuous human habitation since the fifth century BCE. Archaeological explorations provide a better understanding of the culture of these older civilizations. Unfortunately, many of the dwelling mounds have been completely or partially leveled.

The dwelling mounds were constructed from grass sods, domestic wastes, and animal manure. When this rich mix was added to sandy soils, it improved the fertility and water retaining characteristics of the sand. This process was used in the nineteenth century to make some large upland moors usable for agriculture. Digging up the mounds and selling the soil was very profitable, even if homes were destroyed in the process. One hectare (2.5 acre) of sandy upland soil required approximately 80 cubic meters (2,800 cubic feet) of dwelling mound soil fill to make it usable for agriculture (Vollmer et al. 2001).

Dwelling mounds were destroyed on a large scale. This occurred primarily in the period between 1840 and 1945, peaking around 1920 (Meijer 1996). Of the more than 1,200 dwelling mounds, over half were leveled and virtually none were left undamaged. The earth was moved by barges. These barges had easy access to many of the dwelling mounds by way of the dense system of canals that had been dug by that time. In some cases, when part of a mound was leveled, the houses were disassembled and then rebuilt after the earth was removed. In many cases, though, the structures were simply torn down and not rebuilt. Often only the church and the graveyard would be left standing. There were some cases in which the digging was so close to the church graveyard that bones would protrude from the excavation wall. In the process of excavation a great deal of archaeological material was lost. **Figure 2-4** shows the excavation operation in the town of Eenum in 1925. **Figure 2-5** shows how the Frisian dwelling mound, Hogebeintum, was excavated. Most of the area within the elevated ring road was removed, leaving the church and a few homes at the top. **Figure 2-6** is a recent photograph of this dwelling mound. As mentioned earlier, Hogebeintum is the tallest dwelling mound in the Netherlands. Several homes still exist along its ring road.

Archaeological Explorations in Ezinge

The leveling of many dwelling mounds meant a loss of a great volume of archaeological information. At the beginning of the twentieth century some effort was made to learn about the civilizations that occupied these locations. The first attempts were made at Hogebeintum and Ferwerd in Friesland. The researchers began by making a record of the horizontal distribution of the archaeological

Figure 2-6: Hogebeintum—the tallest dwelling mound in the Netherlands.

finds at these locations. The archaeologist A. E. van Giffen made extensive archaeological investigations at the Groningen town Ezinge between 1925 and 1934. He found that this dwelling mound village was occupied from 300 BCE to the present. The initial settlement was placed directly on the salt marsh surface (see **Figure 2-1**) and was about 30 meters (98 feet) in diameter. It contained two dwellings and a barn raised on piles to protect the harvest from the damp conditions (Lambert 1971).

In the next recognized stage of development, it grew to 35 meters (115 feet), reached a height of 1.2 meters (3.9 feet), and included four farms situated close to each other for protection from the harsh environment. As the dwelling mound was raised further, the farm buildings were rebuilt. The third phase puts the dwelling mound at 100 meters (328 feet) in diameter and 2.1 meters (6.9 feet) high. During the Roman period, it was raised again to a height of 3.4 meters (11 feet) and a diameter of 150 meters (490 feet). It was still occupied by four farmsteads set around a central open space. The walls of the homes were made of woven birch boughs plastered with clay and cow dung. This phase of settlement came to an abrupt end after a fire in the third century CE.

Figure 2-7 is a photograph taken in 1932 showing the archaeological excavations underway in Ezinge.

Figure 2-7: Archaeological excavation of the dwelling mound Ezinge in 1932.

From Groningen Institute of Archaeology (GAI); formerly known as BAI.

Places to Visit

Chapters 2 through 8 each have a final "Places to Visit" section. The intent of this section is to provide additional information to those who are planning a trip to the Netherlands and who would like to further explore some of the things described in this text. The places described here are primarily museums, visitor's centers, and viewing locations for sites of interest. Descriptions will be kept to a minimum. Adequate references will be made to allow the reader to locate more information through the Internet. Address, phone, and opening details will be omitted if this information is available on the Internet in English. Chapter 10 organizes most of the "Places to Visit" listed throughout the book into six excursions designed for a visitor to the Netherlands. The Chapter 10 excursion name is given for each "Place to Visit" after the heading "*Along excursion.*"

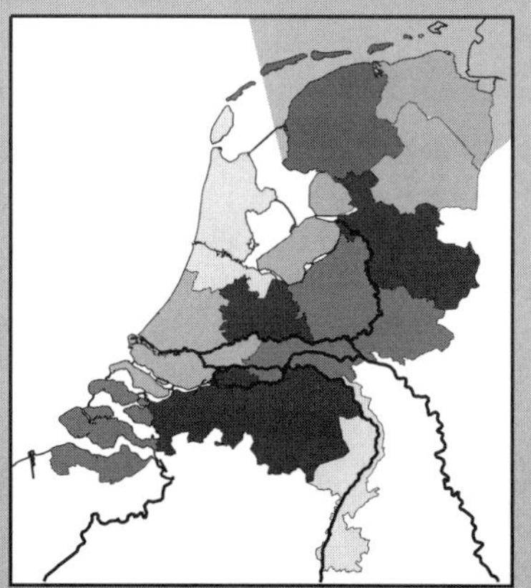

Hogebeintum

DESCRIPTION: Hogebeintum is the highest dwelling mound in the Netherlands. Early in the twentieth century this dwelling mound was nearly leveled for its rich soil. All that remain are the twelfth century church, a few homes on the crest, and several homes along the ring road. The visitor's center provides tours of the church and the dwelling mound. There is limited information in English.

PHONE: 0-51-8411783 (from the United States 011-31-51-8411783).

ADDRESS: Pypkedyk 4, Hogebeintum.

LOCATION: Just east of the town of Ferwerd, which is located along N357 north of Leeuwarden in the province of Friesland.

HOURS: April 1 to October 31: Tuesday through Saturday, 10:00 a.m. to 5:00 p.m.; Sundays and public holidays, noon to 5:00 p.m. November 1 through March 31: Tuesday through Saturday, 1:00 to 5:00 p.m.

ENTRY FEE: Visitor center is free. There is a € 2.50 fee for the guided tour.

INTERNET: No known web site as of April 2005.

ALONG EXCURSION: Frisian Coast and Dwelling Mounds.

Museum Wierdenland

DESCRIPTION: Museum Wierdenland is located in the Groningen town of Ezinge. This dwelling mound village has seen 2,600 years of civilization. The museum highlights the archaeological work done at Ezinge early in the twentieth century. Tours of the church are available. (Contact the museum for tour information.) There is limited information in English.

LOCATION: In the town of Ezinge, which is located about 9 kilometers (5.6 miles) north of the town of Aduard. Aduard is west of the city of Groningen along N355.

INTERNET: www.wierdenland.nl (with some information in English).

EMAIL ADDRESS: info@wierdenland.nl.

ALONG EXCURSION: Frisian Coast and Dwelling Mounds.

3 Draining Noord-Holland Lakes

The sixteenth and seventeenth centuries were a time when the lake draining technologies were developed and implemented on a large scale. For the first time significant gains could be made in the effort to reclaim land that had been lost. The most significant reclamation efforts took place in Noord-Holland north of Amsterdam. A number of large lakes had formed in this region during the twelfth and thirteenth centuries due to drainage of the peat. These lakes were drained and turned into usable agricultural lands. Lake draining at this scale was possible due to the prosperity of the seventeenth century and was necessary due to the growing demand for food.

Loss of Land through 1600

The process of reclamation of peat lands is described in Chapter 1. As the peat was drained, it decomposed and consolidated, causing land subsidence. As a result, the drainage ditches had to be dug deeper to keep the land dry and usable. Eventually the land subsided to the point where gravity alone was not sufficient for drainage, and the land reverted to meadowland. With time, lakes formed at these locations. These lakes formed as a result of human activities in the peat lands, and human intervention was needed to return these lakes to usable farmland.

Figure 3-1 shows a map of the Noord-Holland lakes region around the year 1300. At this time the large tidal estuary, the Zuiderzee, had fully developed, including an arm, called the IJ, which extends just north of Amsterdam. The lakes had grown to the point where this region had as much water as dry land. The inhabitants of this region had to protect themselves on two fronts. Storms from the North Sea came in from the mouth of the Zijpe (near the town of Petten), and the waters of the Zuiderzee posed a threat from the east. By 1300 a large part of this region was protected by the West Frisian Sea Dike (shown in **Figure 3-1**). Other similar large dike rings were added by 1400. The photo in **Figure 3-2** shows one of these dikes along the coast of the former Zuiderzee. These protective dikes were not completely effective. Large storms could still break through on occasion, causing severe flooding. Despite these dangers, the greatest threat came from within the region. The lakes were growing. With each new storm, waves breaking along the unconsolidated peat shorelines caused erosion and loss of land. The residents of the area referred to this devouring action of the wind-driven shoreline erosion as the *Water Wolf* (TeBrake 2002). Because the prevailing wind direction was from the southwest, the new erosion occurred in the northeast parts of the lakes due to wind-driven circulation and wave action. This can be seen in the southwest to northeast alignment of the lakes in **Figure 3-1**. The lakes were also deepest in the northeast corner. The largest of the lakes in this region were the Purmer, Beemster, Schermer, and Heerhugowaard. With time these lakes would have coalesced into one and eventually would have flooded much of this region.

▸ **Figure 3-1:** Noord-Holland around 1300. Prior to the reclamations of the seventeenth century, lakes dominated this area.

Redrawn from van Duin and de Kaste 1990, Fig. 7.

▾ **Figure 3-2:** Dike following the former Zuiderzee coast.

Early Attempts at Lake Drainage

It was clear that the lakes in Noord-Holland were dangerous. The most obvious way to eliminate the erosion of the shoreline was to remove the water. As a result, lake draining became the method of choice for protecting Noord-Holland from this danger within. Lake draining also had another important benefit. It created new land for development. Much of the peat topsoil had decomposed in the initial draining of the land, leaving lake bottoms that were composed of very fertile clayey soils.

The first lakes were drained in the sixteenth century. The smaller lakes were drained first. To successfully drain a lake, those in charge of the project had to follow several steps. First, they cut off any connections to and from the lake and redirected existing streams around the lake. Next, they dug a ring canal around the perimeter of the lake and used the spoils from the canal to build a dike between the canal and the lake water. In some places, low areas on the outer side of the canal were also filled. Next, an outlet for the ring canal was found. Outlet canals were usually dug connecting the ring canal to a receiving water body—in most cases this was the Zuiderzee. Since the Zuiderzee was directly connected to the North Sea, its water level varied with the tides. The excess lake water could be discharged only at low tide. The engineers therefore designed and built sluices at the Zuiderzee end of the outlet canals. These sluices allowed discharge only when the outlet canal water level was above that of the Zuiderzee. They also designed these outlet canals, also known as *boezems,* with enough storage capacity to hold drained water during extended periods of high Zuiderzee water levels.

Figure 3-3: Ring dike and canal along a drained lake. The old land is to the right of the canal and the drained lake is to the left of the dike road.

Once the ring canal, boezem, and outlet sluice were constructed, the water from the lake was pumped into the ring canal with wind-powered water pumping mills. The smallest lakes required only one windmill. Lakes of 600 hectares (1,500 acres) or more needed at least three (Meijer 1996).

The last step of lake draining was that of digging drainage ditches. These were dug inside the drained lake to convey water from the center to the perimeter of the drained lake. When the dry land was split into individual parcels, the drainage ditches were often used as property lines. As a result, the drained lake was completed with a highly regular, rectangular parcelization pattern.

There were two distinct pumping phases associated with draining a lake. The initial phase was intended to simply pump all of the existing water out of the lake. The maintenance phase started once the lake bottom was dry and continues today. Maintenance pumping removes the water that enters the drained lake from precipitation and groundwater seepage. Many of the Noord-Holland drained lakes are now in their fifth century of maintenance pumping. Draining a lake and maintaining the groundwater level within the lake created a new type of polder—the drained lake polder or in Dutch *Droogmakerijen* (to make dry).

The ability to drain large lakes was dependent on the development of the wind-powered pumping mill. Prior to the first lake draining, windmills had been used to grind corn. The pumping mill design was based on the existing corn mill. The development and technology of wind-powered drainage is covered in the next section.

One of the first lakes drained was the Achtermeer. This was a 35-hectare (86-acre) lake south of the town of Alkmaar. It was drained in 1533 (van de Ven 2004). Others included the 43-hectare (106-acre) Dergmeer in 1542, the 686-hectare (1,700-acre) Egmondermeer in 1564, and the 643-hectare (1,588-acre) Bergermeer in 1564. By the year 1600, 19 lakes and 2,747 hectares (6,790 acres) had been drained almost exclusively in the region of Noord-Holland north of Amsterdam. **Figure 3-3** shows one of these drained lakes today. From right to left in this photo are the old land, the outer dike, the ring canal, the inner dike with road, and the drained lake. Trees were used to strengthen the dikes.

Stevin and Leeghwater

Two of the most important contributors to the lake draining technology of this era were Simon Stevin and Jan Adriaensz Leeghwater. Simon Stevin (1548–1620) was a mathematician, scientist, and engineer who was known for solving practical problems using scientific principles. At the age of 29 he was working in the financial administration of the City of Bruges (now Belgium). By the age of 33 he had moved to Leiden where he began his studies at the University of Leiden. At the age of 35, he published his first book on the topic of compound interest. His many following publications cover a broad range of topics, including mathematics (arithmetic, algebra, and geometry), mechanics, hydrostatics, astronomy, navigation, drainage technology, military science, architecture, bookkeeping, music, and civic matters (Dijksterhuis 1970). He is most famous for creating tables for the calculation of compound interest and bringing decimal fractions into daily use. He was closely connected with Maurits of Orange, Stadholder of the Seven United Netherlands. Stevin tutored Maurits, instructing him in many areas of science and mathematics.

Stevin's technical contributions include work on water mills, drainage sluices, sailing vessels, and fortifications. He was credited with the invention of a wheeled carriage with sails, powered by the wind instead of a team of horses. Around the year 1600 Maurits traveled in this carriage, along with 26 others, on the seashore from Scheveningen to Petten, a distance of about 90 kilometers (56 miles).

Stevin is also known for developing, at the request of Prince Maurits, a curriculum in "Dutch Mathematics" at the University of Leiden. The word "Dutch" denotes that this course of study was taught in the national language instead of Latin—the common language of university study of the day. The course of study was intended to provide university instruction for those interested in more practical pursuits, such as engineering, surveying, and cartography. This tradition of Dutch language instruction designed for those involved in technical professions continued for several centuries. In 1807 a national advisory committee reported to King Louis Napoleon Bonaparte that Latin should continue as the language of instruction except in those areas of civil and military engineering (van de Ven 2004).

Stevin held a number of patents associated with windmill drainage. He was known for making improvements to windmills. His detailed calculations improved the way in which power was transferred from the sails to the water wheel via the wooden cog gears. He improved the design of the water wheel itself by making calculations as to the optimal speed and number of vanes. He is also credited with patenting the idea of using windmills in series to increase lift. This innovation (discussed further in the next section of this chapter) allowed for drainage of deeper lakes.

Jan Adriaensz Leeghwater (1575–1650) is the most well known of the Dutch windmill designers and builders, also known as millwrights. He was born in the town of De Rijp, which is situated between two large Noord-Holland lakes, Schermer and Purmer. He gave himself the name Leeghwater, literally translated "empty water," as an indication of his profession.

Leeghwater started out as a carpenter and millwright. By 1604 he had already developed a windmill for the purpose of crushing oilseed (Lambert 1971). He proved his skills in the construction and improvement of drainage windmills after he was hired to oversee the construction of the windmills in the Beemster Polder (discussed in a later section of this chapter). His skills were also sought for drainage projects outside of the Netherlands. Near the end of his life, he took time to write about his accomplishments. His writings also looked ahead to even larger reclamation efforts.

Windmills Used for Draining Lakes

The key factor in the ability of the Dutch to drain the lakes was the availability of mechanical means to lift water—namely, the water pumping windmill. As mentioned earlier, windmills had been used for many years to mill grain. The windmill was not a Dutch invention. It is presumed that returning crusaders first saw windmills in the Middle East, where they were used to lift water from rivers to fields for irrigation.

By the year 1400, corn mills were in use and had achieved a significant level of sophistication. There were two different types of windmills at that time, the tower mill and the post mill. Both of these were redesigned to pump water.

The tower mill consisted of a round, hexagonal, or octagonal structure with a rotating cap at the top. A rotating wind shaft supporting the sails entered through the cap. As the wind direction changed, the cap was rotated so that the sails were always facing the wind. Power was transferred from the wind shaft to a vertical shaft that ran down the center of the mill by way of gears in the cap. The grinding wheel used to mill the grain was located inside the stationary structure. This type of mill was often built along city ramparts.

The post mill was smaller. The entire mill house was elevated and mounted on a massive vertical post supported by four inclined posts. To keep the sails facing the wind, the entire structure rotated on bearings attached to the vertical post. The elevated entrance to the mill was equipped with a set of stairs that rotated as the mill house rotated. A further distinction involved the area under the mill house. An open post mill was built with an open structure below the mill house, whereas in a closed post mill the supporting structure was enclosed.

These windmills, used primarily to grind corn, were later adapted to pump water. The earliest known documentation of water pumping windmills was the rights granted to the city of Haarlem in 1274 by Count Floris V (Stokhuyzen 1962). The earliest water pumping windmills that developed from the post mill appeared in the fifteenth century. The main challenge of adapting a post mill for water pumping was that the pumping mechanism had to be located below the elevated mill house. To do this, the vertical drive shaft of the mill needed to pass through the middle of the vertical support post. For this to work, this solid post was replaced by a hollow post as shown in **Figure 3-4**. In addition, the size of the cap was reduced since it no longer housed the milling equipment. This type of mill is called a wip mill or hollow post mill. **Figure 3-5** shows one of these mills in the village of Cabauw, which lies near the Lek River halfway between Utrecht and Rotterdam. This mill is named "De Middelste Molen" (translated "The Middle Mill") because it was the middle of a group of three. The first mill at this location was built in 1454. After this mill burned in 1772, the windmill in the photograph was constructed as a replacement in a period of only four months (Erik Stoop, personal communication).

The large windmills used to drain lakes were referred to as *polder mills*. Many of the polder mills were a variant of the tower mill. One of the earliest records of the construction of a large polder mill was in 1526. In the seventeenth century the number of these windmills increased dramatically. The major difference between the wip mill and the polder mill was more than just size. In the wip mill the entire upper half of the structure rotated around the center post to turn the sails into the wind. In the polder mills only the upper cap rotated. These caps were usually thatched and rotated on a large roller bearing. Because only the upper caps rotated, they were given the name *bovenkruiers* (upper winders).

The large polder mills that were built in Noord-Holland were designed with the turning mechanism inside the cap. This required the miller to climb all the way to the top of the windmill every time the wind direction changed. These inside turning windmills are also referred to as *binnenkruiers* (inner winders). As the use of this type of windmill spread throughout the country, the millwrights moved the turning mechanism from the inside of the cap to the outside of the mill. This modified version was known as a *buitenkruier*, or outside winder.

Figure 3-6 is a picture of a binnenkruier located near Schermerhorn in Noord-Holland. Note that the brake mechanism is operated from the outside (see the rope hanging below the brake arm), whereas the turning mechanism is on the interior. **Figure 3-7** shows a buitenkruier at Kinderdijk along the Lek River just east of Rotterdam. Kinderdijk's 19 windmills make it the location with the largest number of operating windmills in the world. In 1997 it was named a UNESCO World Heritage Site. In both **Figures 3-6** and **3-7** the sail cloth is folded tightly against the solid part of the vanes. When the mill is in operation, the sail cloth is unfolded and held in place against the wooden lattice (seen in both photos). This increases the effective area of the vanes and increases the amount of wind power captured by the windmill.

▾ **Figure 3-4:** Wip mill or hollow post mill adapted from the post mill used to grind corn. The upper half of the structure rotates and is supported by the hollow post.

From Harte 1849, in the collection of Vereniging De Hollandsche Molen, Amsterdam. (Society The Dutch Windmill.)

▸ **Figure 3-5:** Hollow Post Mill located in the village of Cabauw.

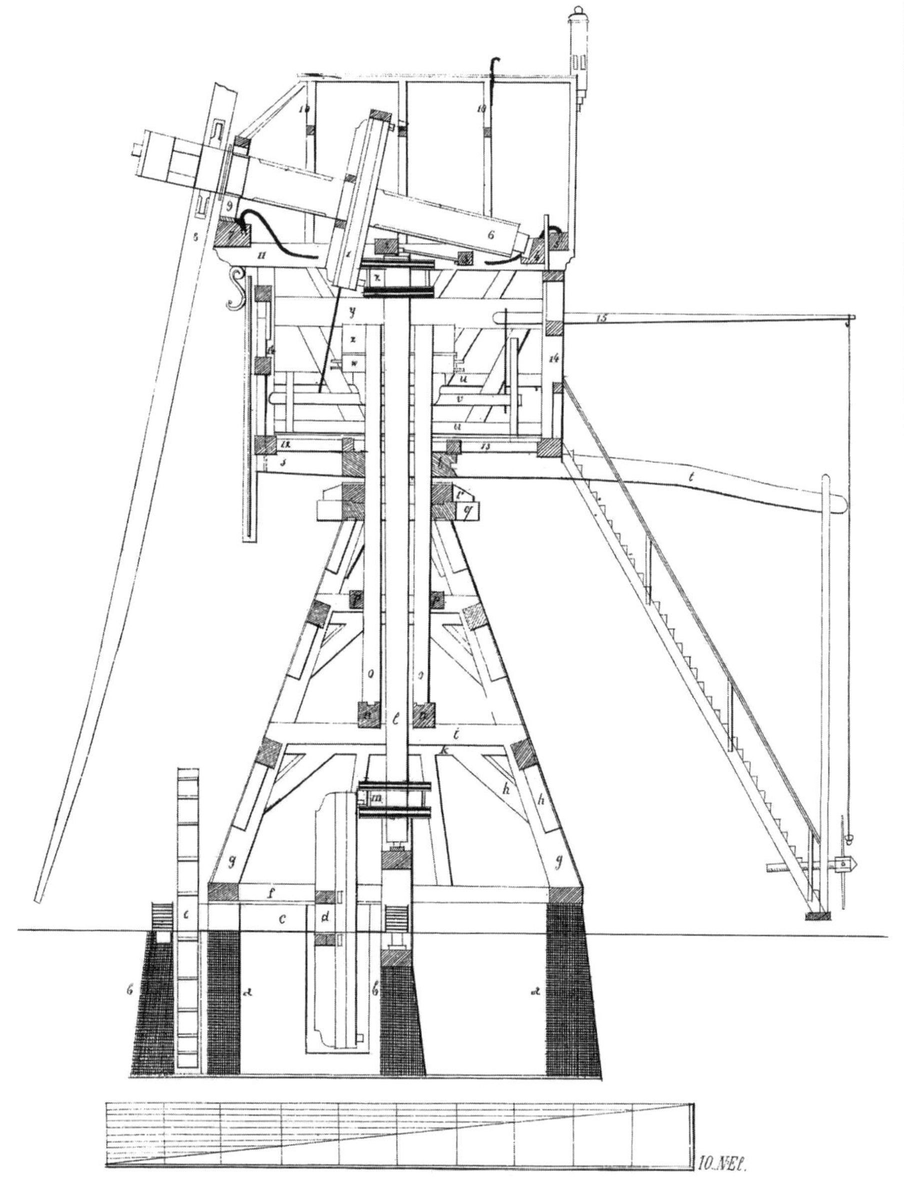

Figure 3-8 shows the internal workings of a buitenkruier polder mill. One of the first things to note is that the vanes do not operate in a vertical plane but are tilted back slightly. The mill was designed this way because it enhances the stability of the structure. It allows the vanes to be larger for a given height structure, thereby capturing more wind. A tapered mill building also allows the windshaft to be shorter.

▾ **Figure 3-6:** Noord-Holland Polder Mill. This windmill located near Schermerhorn has the turning mechanism completely in the cap. The brake is operated by the pole and rope hanging out of the back.

⯯ **Figure 3-7:** Zuid-Holland Polder Mill located at Kinderdijk. The external frame is used to turn the cap into the wind.

◢ **Figure 3-8:** The internal workings of a buitenkruier polder mill.

From Harte 1849, in the collection of Vereniging De Hollandsche Molen, Amsterdam. (Society The Dutch Windmill.)

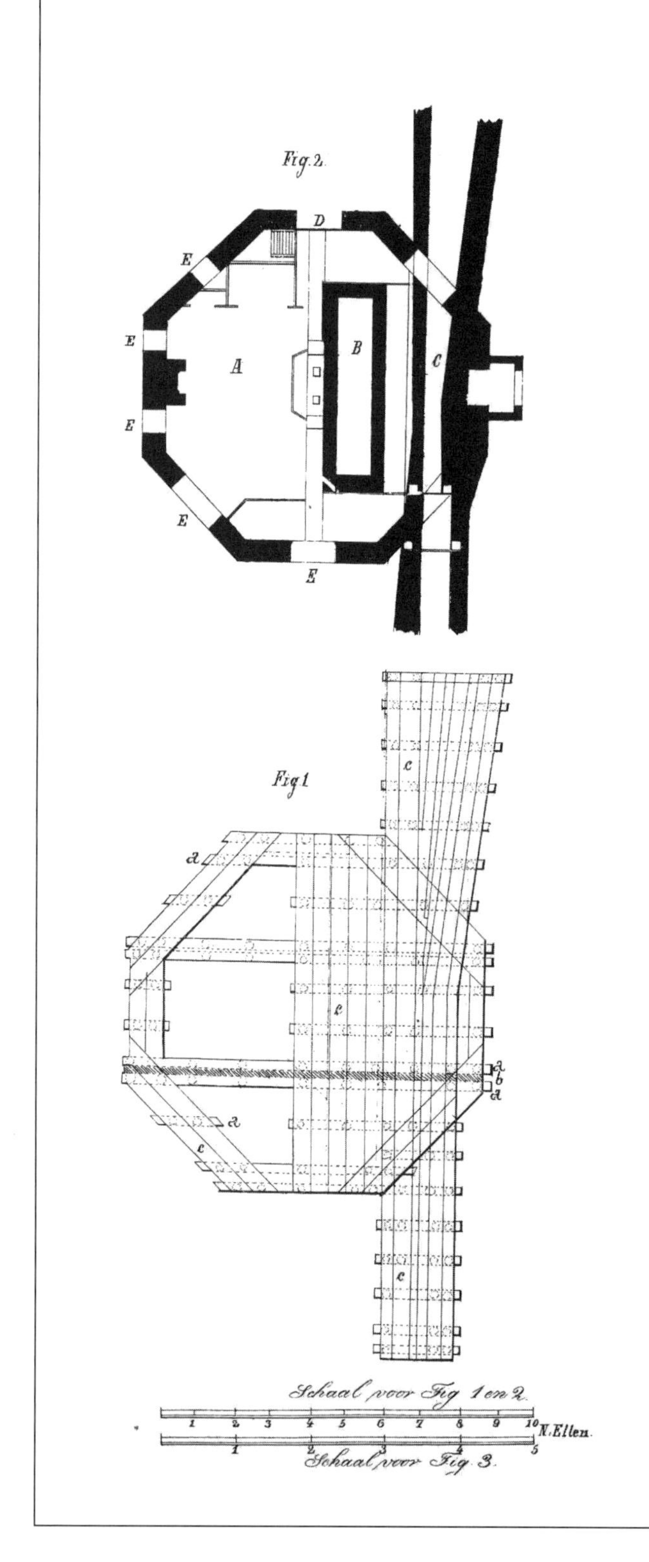

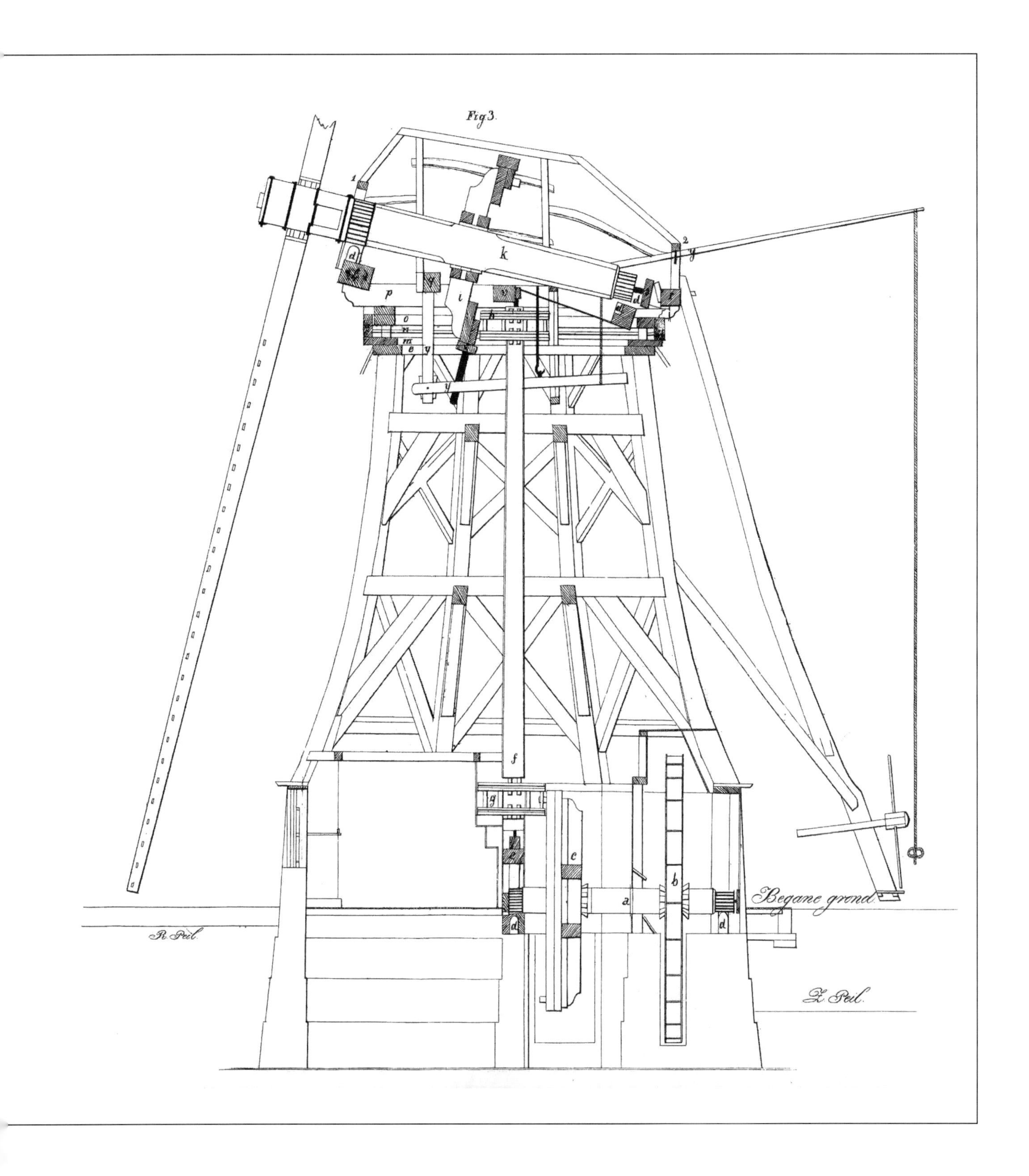
Fig 3.
k
p
i
f
a
b
c
d
Begane grond
R. Peil
Z. Peil

▾ **Figure 3-9:** An Archimedean Screw used in a polder mill.

From the collection of Vereniging De Hollandsche Molen, Amsterdam. (Society The Dutch Windmill.)

▸ **Figure 3-10:** An Archimedean screw used in a polder mill.

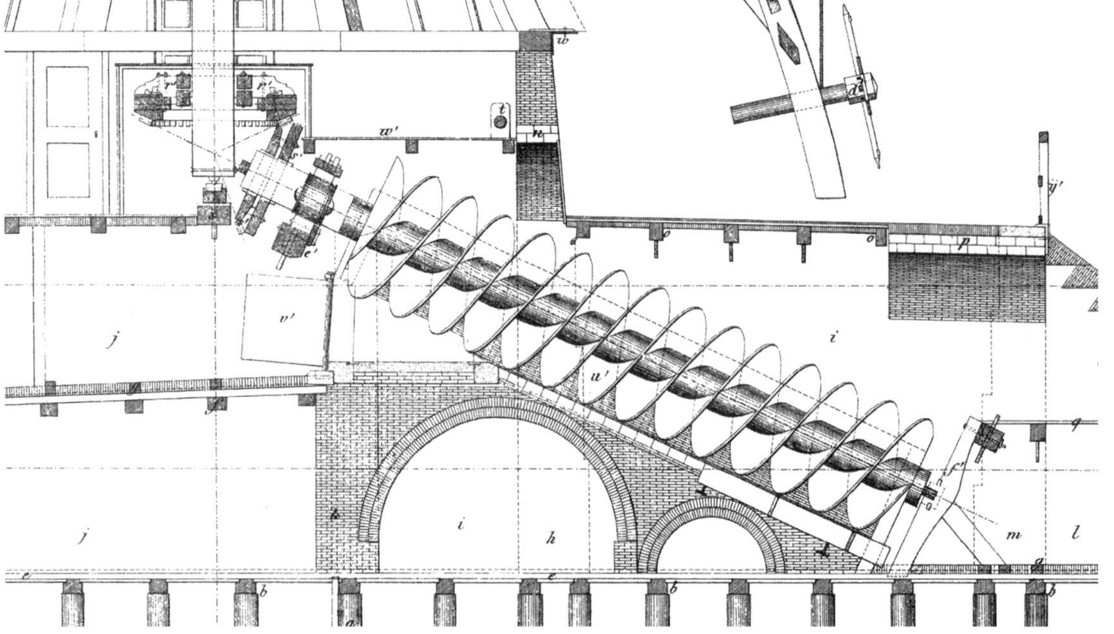

The earliest windmills lifted water using a scoop wheel as shown in **Figure 3-8** (noted with a "b"). It was not until the seventeenth century that these were replaced with Archimedean screws as seen in **Figure 3-9**. Windmills equipped with a scoop wheel could raise water a maximum of 1.5 to 2 meters (5 to 7 feet). The main advantage of replacing the scoop wheel with the screw was that the mill could then lift the water as high as 4 to 5 meters (13 to 16 feet) (van de Ven 2004). The Archimedes screw was first used at Starnmeer in 1643. This modification was so much more effective that a mill with a screw could often replace up to three scoop wheel mills. By the nineteenth century nearly all of the existing windmills were converted. A typical screw pump was triple threaded, had a pitch of 4 to 5 meters (13 to 16 feet), a diameter of 1.5 to 2.1 meters (4.9 to 6.9 feet), and an axle diameter of 0.4 to 0.5 meters (1.3 to 1.6 feet). One of these screws is shown in **Figure 3-10**.

Inside the cap of a polder mill were several important parts of the windmill mechanism. The brake wheel and the bearing used to support the windshaft were both located under the cap. The brake wheel was attached to the windshaft and was activated from the exterior. **Figure 3-11** shows a typical brake mechanism in greater detail. By pulling the brake rope down and to the side, the brake lever is disengaged from the brake catch and is allowed to fall by its own weight. This creates tension on the brake band and causes the brake blocks to stop the windshaft. Pulling the brake rope down again lifts the brake lever to the brake catch. The brake lever is then secured, allowing the windshaft to move again. The brake wheel is also the gear that turns the vertical shaft. The gear teeth are constructed of wood. **Figure 3-12** shows the brake band and wooden gears on one of these brake wheels. The vertical drive shaft is connected to another set of gears in the lower part of the mill. These gears transfer the power from the vertical shaft to the scoop wheel or Archimedean screw.

The capstan wheel used to turn the cap of one of these buitenkruiers is shown in **Figure 3-13**. One end of a chain is hooked around a fixed anchor, in the direction that the cap is to be moved, while the other is wrapped around the shaft supporting the capstan wheel. The miller then turns the wheel to pull the cap around. **Figure 3-14** shows the bearings on which the cap rotates.

Figure 3-11: Detail of a windmill brake mechanism.

Reprinted, by permission of Vereniging De Hollandsche Molen, Amsterdam (Society The Dutch Windmill), from Stokhuyzen 1962.

Figure 3-12: Brake band and gears on brake wheel.

Figure 3-13: Capstan wheel of buitenkruier polder mill. To rotate the mill cap the chain is connected to one of the anchor posts and the capstan wheel is turned.

Reprinted, by permission of Vereniging De Hollandsche Molen, Amsterdam (Society The Dutch Windmill), from Stokhuyzen 1962.

Figure 3-14: Roller bearing on which the windmill cap rotates.

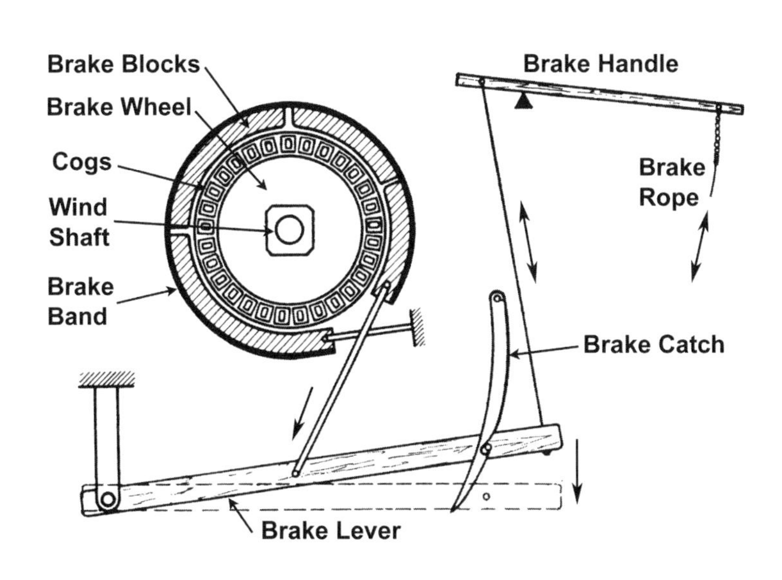

Figure 3-15: Three windmills used in series.

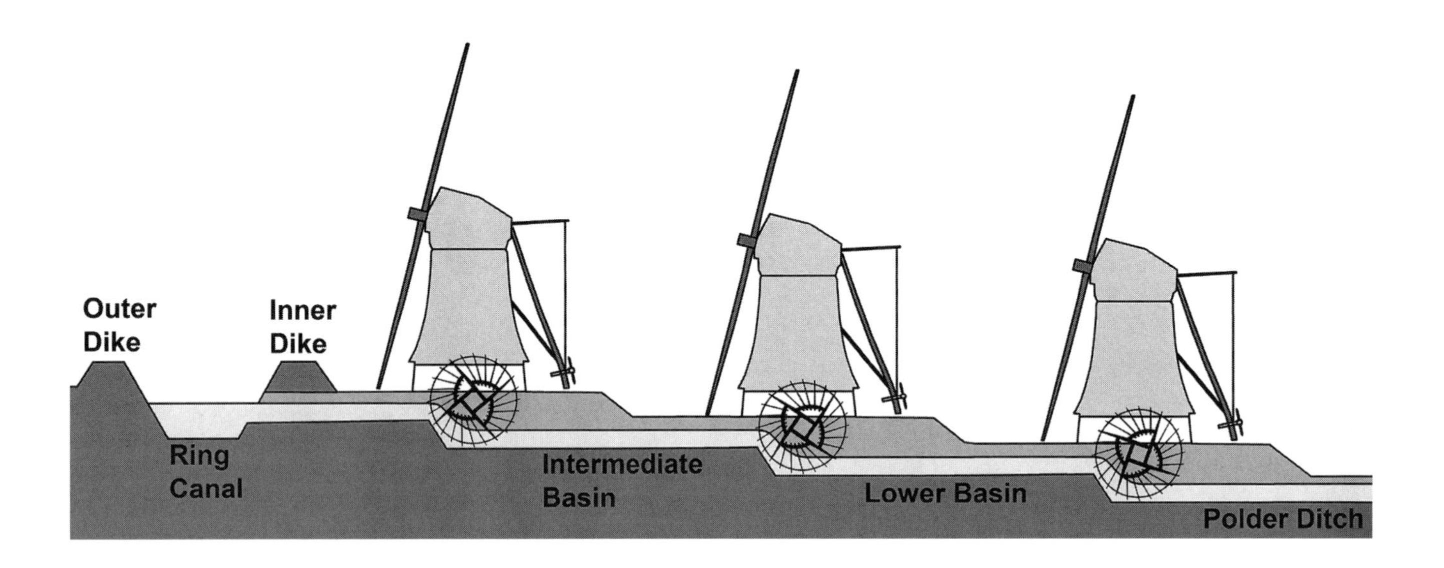

In Noord-Holland the large polder mills had a typical sail span of 25 to 27 meters (82 to 85 feet). They generally operated with wind velocities between 6 and 11 meters per second (20 to 36 feet per second). If the wind speed was too low, the sails would not turn. If the windmill was turning with wind speeds that were too high, friction could set the windmill on fire.

Leeghwater was one of the first to attempt to measure the pumping capacity of a windmill. A later nineteenth century survey of the capacity of 945 Dutch windmills resulted in an estimated capacity equal to 0.9 cubic meters per second (32 cubic feet per second) when lifting the water to a height of 1.0 meter (3.3 feet). This value was a close match to that computed by Leeghwater some 200 years earlier (van de Ven 2004).

There were several technological innovations in the design and use of water pumping windmills. One of the limitations of the sixteenth century windmills was the lift. A typical scoop wheel style pump could lift the water a maximum of 1.5 to 2.0 meters (4.9 to 6.6 feet), operating at a speed of 5 to 8 revolutions per minute. This was sufficient for draining small lakes but not nearly enough to drain some of the larger lakes. Stevin patented the idea of placing windmills in series in which one mill pumps into the inlet channel of the next. See **Figure 3-15**. Some of the lakes drained in the seventeenth century made use of as many as four windmills in series (see the section on seventeenth century projects below).

By the nineteenth century there were about 9,000 windmills in operation in the Netherlands (Stokhuyzen 1962). Early in the twentieth century many of the windmills were removed as steam, diesel, and electric powered pumps were installed. In 1923 the *Dutch Windmill Society* was established with the goal of preserving the old windmills. Today about 1,000 windmills remain, many of which are maintained in good working order.

Seventeenth Century Projects

By the end of the sixteenth century a total of 2,747 hectares (6,790 acres) of lake had been drained (van de Ven 2004). The efforts in the first half of the seventeenth century resulted in a reclaimed area almost ten times the size of that of the sixteenth century—48 lakes and 27,000 hectares (66,700 acres). Several elements combined to create this sudden increase in lake draining activity. Increased population created a greater demand for agricultural products. At this time the country was experiencing a great deal of wealth as a result of international trading activities. The wealthy members of the Dutch East Indies Company needed something in which to invest some of their wealth. Often several merchants would pool their capital and form a company for lake drainage. With this increased capital, much larger projects could be tackled.

There were several reasons why the area north of Amsterdam was a great location to invest in lake drainage. First, the soil in the reclaimed lake bottom was usually the very productive clay that lay beneath the old peat surface. Second, this area was situated very near the large population centers, making it good for growing perishable goods such as vegetables and dairy products. Third, the resulting new countryside was also a good location for the rich Amsterdam merchants to build their country homes.

The first large lake to be drained early in the seventeenth century was the Beemster. At 7,220 hectares (17,800 acres), this lake was 12 times larger than any drained previously, and it was 2.6 times larger than the total area of lake drainage from the entire sixteenth century. Its depth varied between 1.7 and 3.0 meters (5.6 and 9.8 feet), requiring the use of several groups of pumping windmills in series (Meijer 1996).

The company formed to drain the Beemster was led by one of the directors of the Dutch East Indies Company, Dirck van Oss. Van Oss was credited with contracting with Henry Hudson to search for a new route to the Far East. In 1607 a group of 16 investors requested permission from the authorities, and a patent was granted to establish the *Beemstercompagnie*. The patent speaks of "work such, that it is possible to make Water into Land." Negotiations were made with the water board to show that provisions were made to manage the water that used to drain into and out of the lake.

To drain the Beemster, the lake was first hydraulically separated from the rest of the area by constructing a ring dike around the entire lake. The drainage from the surrounding land was sent to the Zuiderzee by an 8-kilometer (5-mile) canal. The ring canal and ring dike around the lake were 50 kilometers (31 miles) long. The initial plans called for 16 windmills to be used in groups (or gangs) of two. The 16 windmills were to first work on their own. When the lake level dropped by 1.25 meters (3.6 feet), half of the mills would have their wheels lowered and the lower mills would then pump to the remaining upper mills. Shovels and pickaxes were used in the basic engineering works; the foundations for sluices and windmills were sunk using manual pile-driving installations operated by 30 to 40 people.

The contract specifications had provisions for miscalculations in the windmill-pumping rate. The specifications read as follows (van de Ven 1994):

> And to make sure that the reclamation can be carried out by means of sixteen windmills for drainage, the contractors have accepted and promised, just as they hereby accept and promise, that they will pass tests of their invention full-size, in the presence of the Delegates of the Landowners in the polder within a period of four weeks at most. And if it is found that the sixteen windmills for drainage are not capable of doing the work, the contractors will be held to provide themselves immediately with more old or new windmills up to the number the Delegates see fit.

The resulting tests showed that 16 windmills would be inadequate. By 1609 an additional ten were constructed, bringing the total to 26. Later two additional pairs were added, bringing the total to 30.

By 1610 the lake draining was nearly completed. Most of the lake bottom was visible, and surveyors were already laying out the drainage ditches. Then disaster struck. A storm broke through the Zuiderzee dikes and completely flooded the nearly dry lake. The owners were faced with a difficult decision. They had to decide whether to start over or to simply count this as a lost investment. Fortunately, they decided to finish what they had started.

They did, however, make some changes. To begin with, they raised the ring dike one meter above the ground level of the old land. They also added more windmills, bringing the total to 40. As the lake began to fall dry once again, the owners discovered that the northeast end was too deep to be drained with gangs of only two windmills. Four more windmills were added, providing four gangs of three mills for the deepest section (now 44 windmills total). Finally, by July 4, 1612 the lake was dry enough to begin building roads.

After the lake was dry, drainage ditches were dug, bringing the water directly to the windmills. These drainage ditches formed the basis of the land parcelization within the drained lake. After the drainage ditches were in place, the land began to subside. Consequently, the drainage ditches had to be deepened. This required four-stage pumping at six locations. Finally, with the

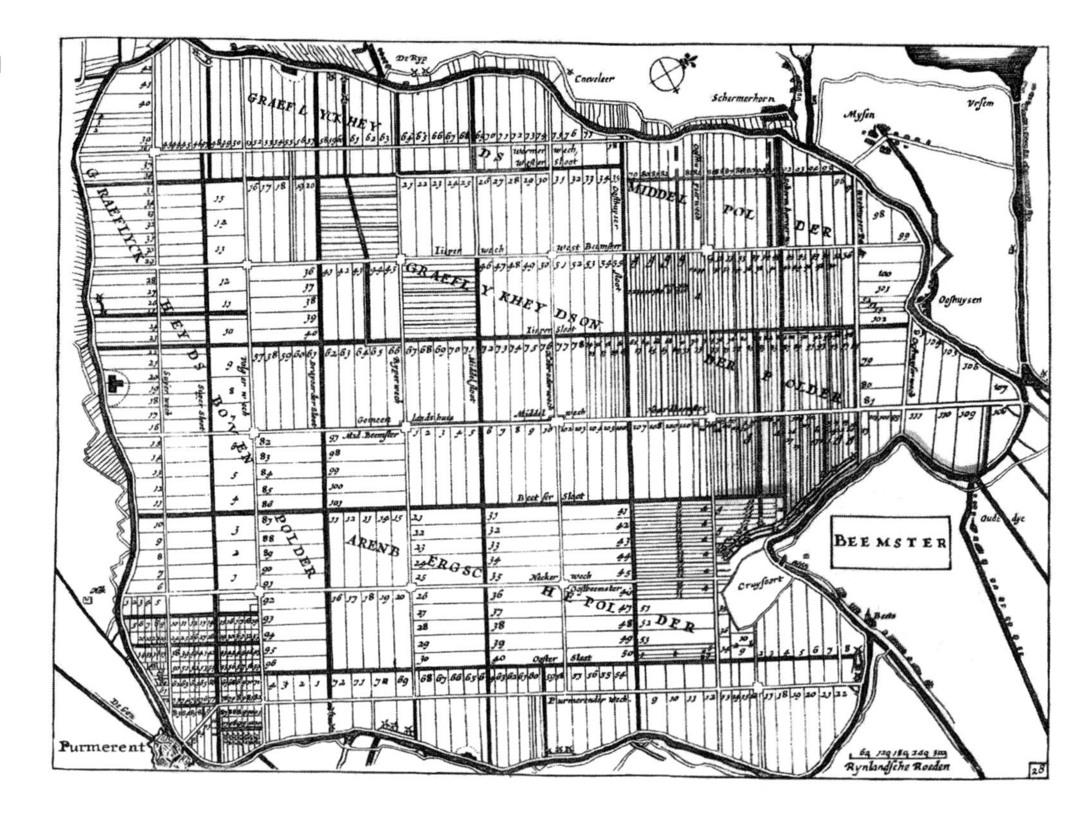

Figure 3-16: Map of the Beemster Polder produced in 1635.
From Colom 1635.

help of 50 windmills, the former Lake Beemster was maintained in a dry condition. This was a far cry from the 16 windmills proposed in the original plans. Lake Beemster finally became the Beemster Polder.

Parts of this polder lay 3 to 4 meters (10 to 13 feet) lower than the surrounding low countryside. Nevertheless, it was considered safer than the surrounding region. In 1625 a flood broke through the seawall along the Zuiderzee. People found that they could escape the flooding by taking refuge in the lower Beemster Polder (van de Ven 2004). This was the ultimate test of the design and construction of this polder.

The windmills continued to operate in the Beemster Polder until 1877 when they were replaced with steam pumps. Unfortunately, all of the old windmills were destroyed, robbing the Beemster of its legacy. Today two diesel pumps have sufficient capacity to keep this area dry.

The parcelization of the Beemster occurred in the process of building drainage ditches within the polder. As was the custom in the seventeenth century, parcelization followed strict rectangular patterns. Most of the parcels were 180-meter-by-900-meter (590-foot-by-2,950-foot) plots. This parcelization is still evident today. **Figure 3-16** shows a map of this area published in 1635 by J. A. Colom.

Former Lake Beemster proved to be a worthwhile project for its investors. By 1640 there were 50 large estates, 200 farmhouses, 150 homes, and 2,000 people whose feet remained dry because of the 50 windmills (Spier 1969). In total there were 123 investors, who received a return of 17 percent on their investment upon completion of the polder in 1612.

In 1999, at the twenty-third meeting of its World Heritage Committee, UNESCO designated the Beemster Polder a World Heritage Site. The Committee referred to the Beemster Polder as a masterpiece of creative planning, in which the seventeenth century ideals can still be seen in the design of the landscape. The distinctive Beemster landscape with its strict rectangular pattern of land division and unusual history is the only one of its kind in the world. By designating it a World Heritage Site, UNESCO officially recognizes the value of the Beemster Polder.

The process of draining lakes in the region north of Amsterdam continued through the first half of the seventeenth century. Other lakes drained included the Purmer at 2,750 hectares (6,800 acres) completed in 1622, the Heerhugowaard at 3,530 hectares (8,720 acres) completed in 1625, the Wijde Wormer at 1,650 hectares (4,080 acres) completed in 1626, and the Schermer at 4,450 hectares (11,000 acres) completed in 1635.

Out of the 86 lakes reclaimed in the Netherlands in the sixteenth and seventeenth centuries, 69 were in Noord-Holland, accounting for 93 percent of the drained area (van de Ven 2004).

Places to Visit

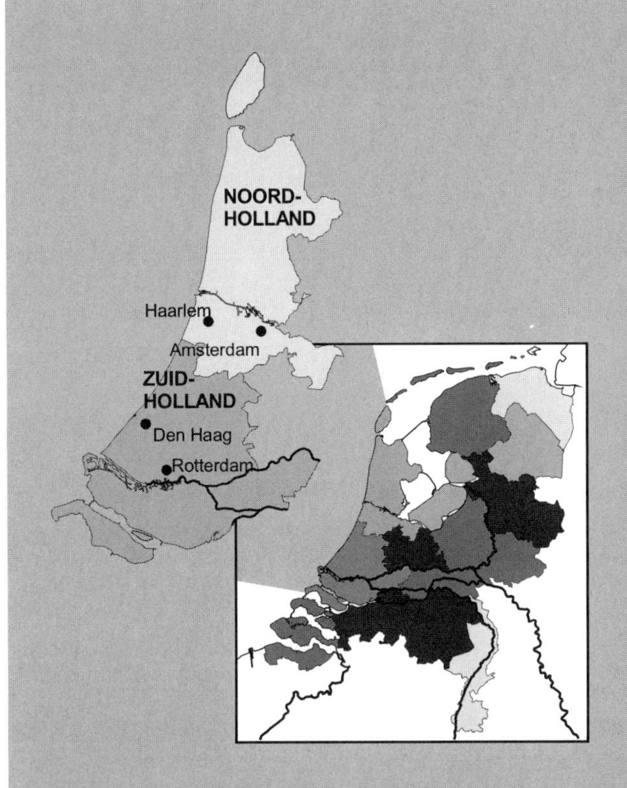

Windmill Museum Schermerhorn

DESCRIPTION: This is a windmill museum in the Schermer Polder. It includes a working polder mill and a small museum. In the museum you can view an English language slide show. You can tour the windmill and climb to the cap to see how it works.

LOCATION: Along N243 just west of the town of Schermerhorn. Schermerhorn is about 8 kilometers (5 miles) east of Alkmaar in the province of Noord-Holland.

INTERNET: www.surf.to/museummolen.

EMAIL ADDRESS: info@museummolen.nl.

ALONG EXCURSION: Noord-Holland Drained Lakes.

Kinderdijk

See description listed at the end of Chapter 6.

4 Draining the Haarlemmermeer

This chapter considers the drainage and reclamation of the Haarlemmermeer or Lake Haarlem in the middle of the nineteenth century. There are several elements that distinguish the reclamation of the Haarlemmermeer from the seventeenth century lake reclamations north of Amsterdam described in Chapter 3. The most significant difference is the size of the lake. The largest lake drained in the region just north of Amsterdam was the Beemster at 72.2 square kilometers (27.9 square miles). At the time it was reclaimed, the Haarlemmermeer had grown to 180 square kilometers (70 square miles). Earlier proposals to drain this lake by wind power probably could have been successful, but they would have been very expensive. It was not until the arrival of steam-driven pumps that the Haarlemmermeer could be successfully drained.

Another significant difference between the reclamation of the Haarlemmermeer and the seventeenth century lake reclamations north of Amsterdam was that the earlier reclamations were all privately funded. By the middle of the nineteenth century the venture capital needed to take on such large projects was no longer available. As a result, the reclamation of the Haarlemmermeer was a public works project. It was attempted as the country was trying to recover from a less prosperous period and undertaken at a time when the country was beginning to exert its independence as a nation.

Growth and Expansion of the Haarlemmermeer

The Haarlemmermeer grew to its nineteenth century size because of a combination of land drainage, peat cutting, and shoreline erosion during storm events. As discussed earlier, drainage produced decay and consolidation of the peaty soils, resulting in land subsidence. The oldest drained areas turned into waterlogged meadows. As a result, a number of small lakes were formed. Winds pushed waves up on the shores of these peat-surrounded lakes, causing shoreline erosion, which, in turn, resulted in an increase in lake size. **Figure 4-1** shows the growth of the Haarlemmermeer over the period from 1250 to 1848.

Peat cutting was also a significant factor in the growth of the Haarlemmermeer. By the late Middle Ages, the Dutch had started using peat as fuel. There was a great demand for fuel for both domestic and industrial purposes. In a region where the yearly average temperature is only 9° Celsius (48° Fahrenheit), fuel for heat was important. Fuel was also needed for industrial uses—in particular pottery making, breweries, bakeries, and textile manufacturing.

Peat was removed using one of two different techniques. Dry peat cutting methods were employed in higher locations where canals could be dug to lower the groundwater levels below that of the peat. Wet peat cutting or peat dredging was used for peat extraction below

Figure 4-1: Growth of the Haarlemmermeer. The Haarlemmermeer was formed from the coalescence of three smaller lakes—Leidschemeer, Oude Haarlemmermeer, and Spieringmeer.
Redrawn from SWAVN 1986, Fig. 23.

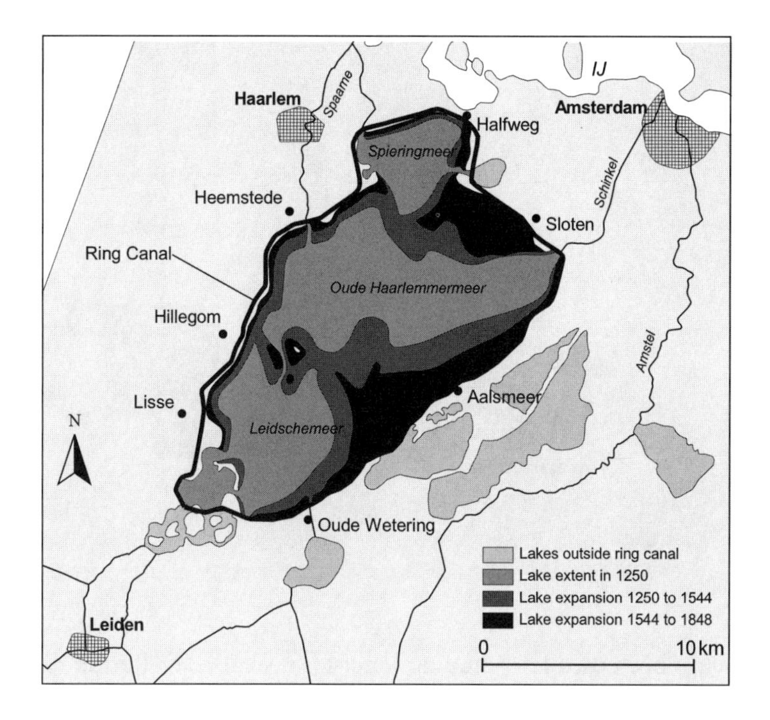

the water table. Dry peat cutting was economical only on a large scale. Canals were constructed to lower the groundwater level and to allow the harvested peat to be transported by boat. After the peat was cut from a site, the land that remained could be turned into productive agricultural land. In these higher locations, the sandy soils at the base of the peat were mixed with the *bonk* or upper layer of peat that was not useful for fuel. With enough additional fertilizer (manure, compost, and street sweepings delivered on returning peat barges) the land could be returned to productive use.

In the sixteenth century, peat dredging techniques were developed to extend the peat harvest to areas where it was not as easy to lower the water table. The peat was dredged from boats using hand drag tools. It was then placed on small strips of ground to be compacted and dried before being cut and shipped away for use. Peat dredging often led to permanent destruction of the land. After an area was harvested, all that remained were artificial lakes or *plassen* with small strips of land in between. Many of these lakes coalesced to form larger lakes. If the soil beneath the peat was clayey, the lake might be drained and reclaimed for agricultural uses. In other areas sandy soil layers lay beneath the peat. Since the sandy soils were not good for agricultural uses, these areas were often left unreclaimed.

Peat cutting and dredging boomed in the seventeenth century. The availability of inexpensive fuel contributed to the economic prosperity of this period. Peat harvesting also provided work for many. The main problem associated with peat harvesting was significant loss of land. In 1530 the Haarlemmermeer was 26 square kilometers (10 square miles) in size. Aided by peat cutting and storm gales, by 1600 it had coalesced with several other lakes to become a much larger lake of 106 square kilometers (41 square miles). By 1700 it grew to 160 square kilometers (62 square miles) and became a serious flooding menace. Efforts to stop or control peat cutting began late in the seventeenth century. Some laws were enacted requiring that all areas subject to peat

▼ **Figure 4-2:** Simple steam-driven pump using a beam to transfer power from the steam engine to the pump. Low-pressure steam fills the cylinder as the piston is drawn upward by the weight of the pump piston.

◀ **Figure 4-3:** Working stroke of an atmospheric steam-driven pump. The steam is condensed in the cylinder, and the pressure difference between the atmosphere above and the vacuum below drives the piston downward.

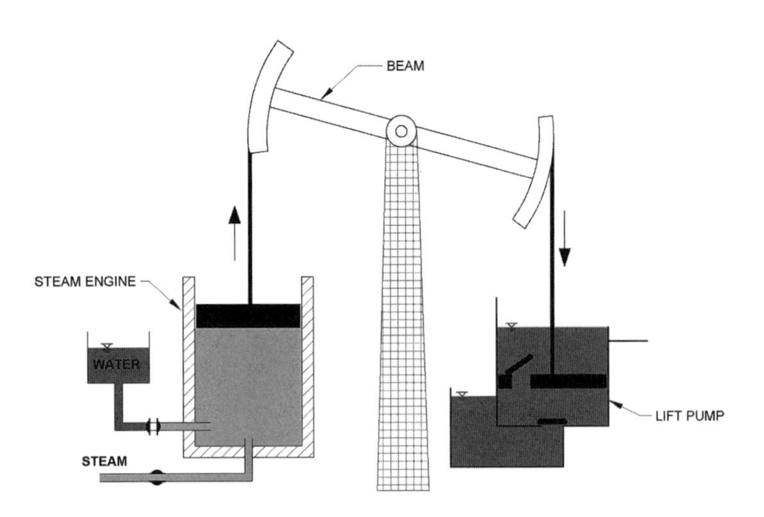

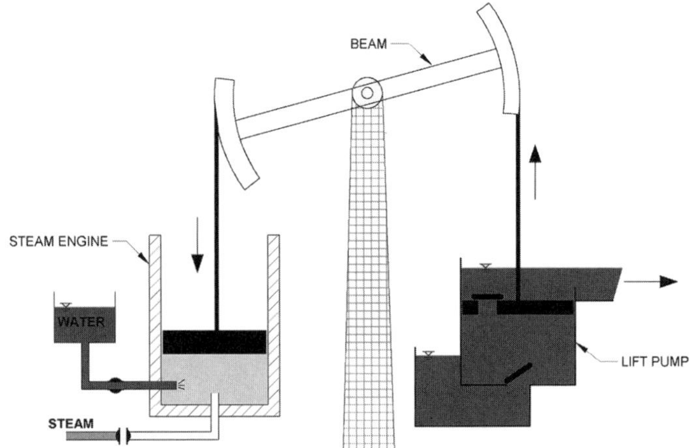

harvesting be reclaimed. Nevertheless, peat harvesting continued until well into the nineteenth century.

The success of lake drainage in the area north of Amsterdam inspired many plans for the draining of the Haarlemmermeer. Over the years literally hundreds of proposals were suggested. One of the earliest was that of Jan Adriaensz Leeghwater, who, in 1641, proposed draining the Haarlemmermeer using 160 windmills. Other plans were proposed by the eighteenth century surveyor Nicolaas Cruquius (1678–1754) and in the nineteenth century by F. G. Baron van Lynden van Hemmen (1761–1845) (van der Pols and Verbruggen 1996). Lynden was the first to propose drainage by steam-driven pumps alone.

Development of Steam Power for Draining Lakes

As previously discussed, wind power was first used in the Netherlands for pumping water in 1408 and for draining lakes as early as 1533. By the middle of the nineteenth century, when the country was ready to embark on draining the Haarlemmermeer, windmills for water pumping had a 450-year track record of success. Steam power was by no means the obvious choice for draining the Haarlemmermeer. At the time, steam power had been used for pumping water for several years, but up to that point it had variable success in the Netherlands.

As a method to pump water, steam power was very different from wind power. The early steam engines had reciprocating vertical shafts. These were used to drive piston pumps instead of scoop wheels or screws. These early steam pumps were referred to as beam engines because the reciprocating motion of the steam engine was transferred to the pump by way of a pivoting beam.

Several different types of steam engines were developed in the eighteenth and nineteenth centuries. One of the first steam engines used for pumping water was the Newcomen engine, developed around 1710 in England for draining mines. The operating principle is shown in **Figures 4-2** and **4-3**. In **Figure 4-2**, steam at nearly atmospheric pressure entered the cylinder of the engine. The piston in the engine was raised primarily from the weight of the pump piston. At the top of the stroke, the steam valve was closed and a small spray of water was injected into the engine. This caused the steam to condense, creating vacuum conditions in the steam cylinder. The unbalanced pressure drove the piston in the steam cylinder downward (see **Figure 4-3**). This was the working stroke. It delivered the water to a higher level.

Because the cylinder walls in a Newcomen engine were cooled and reheated with every cycle, the efficiency

▾ **Figure 4-4:** Watt Steam Engine. The steam is condensed outside the cylinder. The steam jacket also keeps the cylinder hot.

◢ **Figure 4-5:** Sims Annular Compound Engine. This engine uses two concentric pistons.

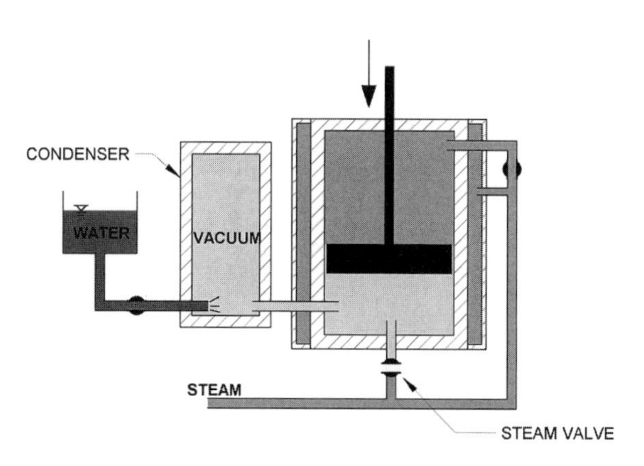

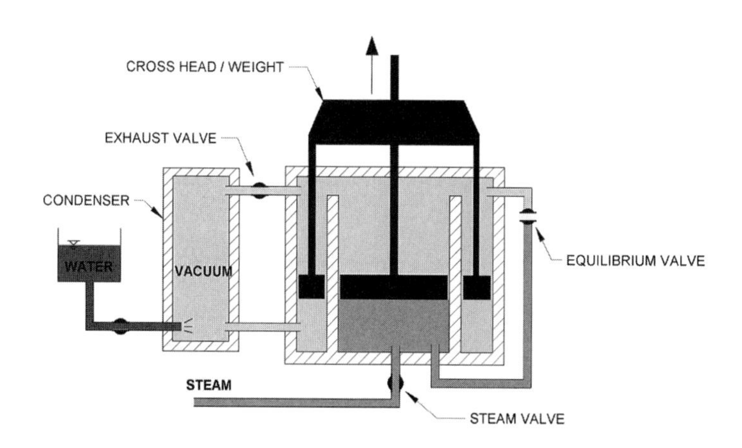

was very low. The efficiency is the portion of the energy supplied that is effectively used to pump water (energy supplied to the water divided by the chemical energy in the fuel). The Newcomen engine's efficiency was between 0.5 and 0.8 percent (van der Pols and Verbruggen 1996). The Newcomen engine was frequently called a *fire engine* in reference to its source of power.

In 1769 James Watt made several improvements to the design of the steam engine. His design is shown in **Figure 4-4**. A separate condenser was added to avoid the heating and cooling cycles in the cylinder. The cylinder top was enclosed and a steam jacket was added, surrounding the cylinder and keeping it hot. Steam (still at virtually atmospheric pressure) was also introduced above the piston on the downward stroke. **Figure 4-4** shows the engine during the working stroke in which the piston was drawn down against the weight of the pump and the water was raised. Efficiencies improved dramatically, rising as high as 2.5 percent (van der Pols and Verbruggen 1996).

The Cornish engine was an improvement on the Watt engine. The first one was built in 1812 by Richard Trevithick after the patent on the Watt engine expired. The primary difference between the Watt engine and the Cornish engine was the Cornish engine's use of high-pressure steam—between 3 and 5 bar (44–73 pounds per square inch). The steam valve in this engine closed after 10 to 30 percent of the stroke. This allowed the expanding steam to provide additional work. With additional improvements, by 1840, this engine achieved an efficiency of 6 to 7 percent (van der Pols and Verbruggen 1996). The main drawback was a "kick" that occurred when the high-pressure steam was introduced.

James Sims reduced the kick problem with the design of a compound engine. This engine featured two pistons with different diameters working together in a single cylinder. A modification of this design is shown in **Figure 4-5**. This engine had a low-pressure cylinder surrounding a high-pressure cylinder. The concentric pistons were connected to a single crosshead that also provided weight for the return stroke. This was the Inverted Sims Annular Compound engine.

At the start of the cycle, the pistons were in the lowest position. The steam valve was opened to begin the upward stroke. At the same time, the exhaust valve was opened, allowing the steam from the previous stroke to enter the condenser where it was condensed by a spray of cold water. This formed a vacuum in the cylinder above the pistons, drawing them upward. The steam valve closed partway through the upward stroke, allowing the pistons to slow to a stop as the high-pressure steam expanded. After a short pause to allow the valves in the

Figure 4-6: Four different reciprocating pumps.
Redrawn from van der Pols and Verbruggen 1996, p. 26.

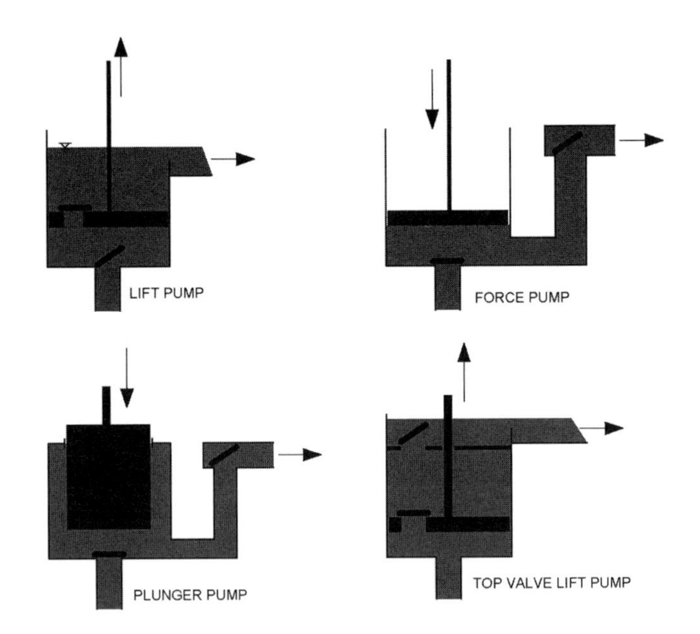

pump buckets to close, the exhaust valve was closed and the equilibrium valve was opened. This allowed the expanded steam to fill the space above the pistons. The pistons now dropped under the force of gravity, assisted by the pressure difference across the outer piston. (The space below the outer piston was still under vacuum conditions.) This type of engine was selected to drive the pumps used to drain the Haarlemmermeer.

Several different types of pumps were used with these early steam engines. The early pumps all used pistons because transferring power from a reciprocating engine to a piston pump could be easily accomplished with a simple beam. Rotative pumps (scoop wheel and Archimedean screw) developed later. **Figure 4-6** shows four different arrangements. The *lift pump* and the *force pump* both draw water through a foot valve (on the bottom of the cylinder) on the upward stroke. The lift pump uses the upward stroke as the working stroke in which it lifts the water above the piston to the outlet. On the downward stroke the clack valve (on the piston) opens, allowing water below the piston to move above the piston. The force pump uses the downward stroke as the working stroke pushing water to the outlet. The *plunger pump* is a variation on the force pump. The *top valve lift pump* is a variation of the simple lift pump.

The first attempt in the Netherlands to use steam power to pump water occurred in Rotterdam in the mid-eighteenth century. Steven Hoogendijk was a clock and watchmaker who also supervised the windmill that was used to flush out the city's canals. The city's canals were used as sanitary sewers. In the summer when there was little movement in these canals, the smell was terrible. Unfortunately, the wind needed to run the windmill did not always occur when the smell was the worst. After a meeting with the famous civil engineer, John Smeaton, Hoogendijk convinced the town council to send a representative to England to determine the feasibility of using steam power to flush their canals (van der Pols and Verbruggen 1996). This happened in 1757 without much success. The pumps used for waterworks in England at that time, particularly in the hilly terrain, were designed for lower flow rates and higher lift than that required for drainage in the Netherlands. The Rotterdam representative did not recommend the use of steam. Hoogendijk did not give up. His next approach was to generate support for steam power by creating a *learned society*. Learned societies were established, beginning in the seventeenth century, to promote doing science by observation and experimentation. Members of these societies, among other activities, organized lectures and conducted experiments. In 1769 Hoogendijk established the *Batavian Society of Experimental Philosophy* with the goal of placing new technologies to work in society. Finally, in 1774, with help from other members of the Batavian Society, Hoogendijk was able to get permission to build the first steam-powered water pump in the Netherlands. The pump was located at the sluice where the canals on the east side of Rotterdam flowed into the Rotte River. A pump here would allow the canals to drain when the level in the Rotte was high. It would also supply water in

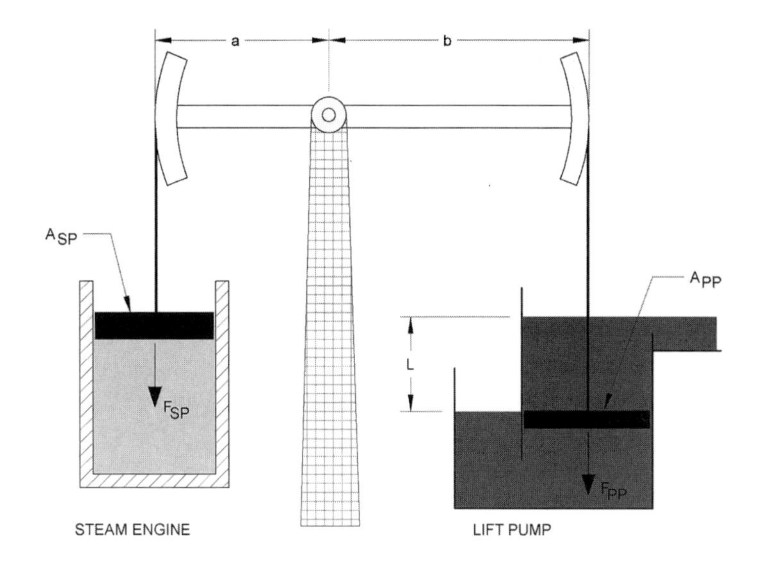

Figure 4-7: Steam engine and pump. The only way to accommodate varying lift conditions was to adjust the pump piston area and the pump pivot distance.

the Rotte when it was low and polluted. To get permission to build the steam pump, Hoogendijk had to fund the project himself.

At this time (1774) Watt was still in the process of developing his engine. He had patented the separate condenser in 1769, but the Watt engine was not ready for the market until around 1775 (Jan Verbruggen, personal communication). The engine selected for this first steam drainage application was a 60-year-old technology, Newcomen engine.

One of the big problems associated with steam-driven pumps is that of varying lift. In the Dutch applications, the height to which the water needed to be raised varied significantly with time. For lake draining, the amount of lift needed increased as the water level in the lake dropped. When using a pump to flush out the canals in Rotterdam, the water surface levels on each side of the pump varied and, as a result, so did the required lift. **Figure 4-7** shows a simplified Newcomen steam engine connected to a lift pump. The steam engine used pressure differences between the two sides of the steam piston to do work. The pressure differences across the piston did not vary much over the range of operating conditions, so a constant force, F_{SP}, was available to do work. The magnitude of this force was $F_{SP}=p_c A_{SP}$, where p_c is the difference in pressure between the atmosphere and steam in the steam cylinder and A_{SP} is the area of the steam piston. In the pump, the force on the pump piston was the pressure difference across the piston times the piston area. The pressure on each side of the pump piston was simply the depth in the connected water times the unit weight of water. The pressure difference was then the unit weight of water times the difference in water surfaces or the lift. So the force needed to lift the pump piston was $F_{PP}=L\gamma A_{PP}$ where L is the lift, γ is the unit weight of water, and A_{PP} is the area of the pump piston. For the pump to be able to lift the water, the moment of the steam piston force about the beam hinge (F_{SP} multiplied by a) had to exceed the moment of the pump piston force about the beam hinge (F_{PP} multiplied by b). This provided a relationship for the maximum lift that was possible given the engine parameters:

$$L < \frac{p_c}{\gamma} \frac{a}{b} \frac{A_{SP}}{A_{PP}}$$

The vacuum pressure and the steam piston area were fixed for a given steam engine. The only way to adjust the pump for varying lift conditions was to change the beam pivot point (i.e., adjust b) or change the pump piston area (A_{PP}). To do this, some of the early pumping stations built for use in the Netherlands had multiple pumps that could be attached to several locations along the beam, thereby allowing varying lift conditions.

The Rotterdam facility had eight pumps powered by a single steam engine and five beams. **Figure 4-8** is a plan view of this pumping station showing eight pumps (four circular and four square) and five beams. Variable lift was made possible by connecting and disconnecting the chains attached to the pump pistons.

This first steam-driven pumping station was completed and put into operation in March 1776. The steam engine worked well, but the beam and pump arrangement failed. Several modifications had to be made to get it to work at all. As a result of these modifications, two of the pump cylinders were taken out of service and two beams were removed. The Rotterdam fire engine eventually

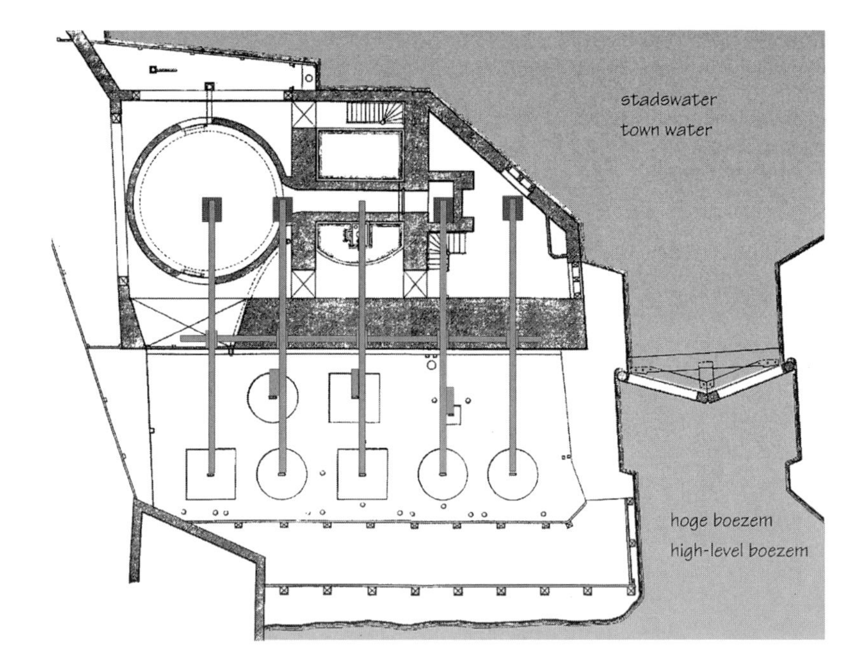

Figure 4-8: Plan view of the Rotterdam fire engine. It shows the arrangement of the eight pumps, all driven by a single Newcomen engine via five beams.

Reprinted, by permission of Museum De Cruquius, from van der Pols and Verbruggen 1996.

failed due to problems with the power transfer mechanisms. The beams and beam supports could not handle both the torsion and bending loads that were applied. The pumping station also failed because wooden pumps were used (Jan Verbruggen, personal communication). The steam engine itself worked well. The facility, shown in the painting in **Figure 4-9**, was eventually abandoned in 1785.

Over the next several decades, more attempts were made at bringing steam-driven pumps to the Netherlands. The advocates of steam-driven pumps had a difficult time arguing for their use. After all, steam pumps were unreliable, inefficient, and used costly fuel. The well-established wind technology had a long track record of reliable operation using free fuel. In comparison, the windmill's limited pumping capacity, limited lift, and unreliable fuel source seemed to be an acceptable drawback. Nevertheless, several other attempts were made at steam-driven pumping in the Netherlands.

In 1781 the banker John Hope wanted a pump for his water garden in Heemstede (near Haarlem) near the dunes. This was a situation in which the lift was very constant. Rinze Lieuwe Brouwer designed a Newcomen engine with a single pump as shown in **Figure 4-10**. Built in the Netherlands, the pump operated well for many years—possibly until around 1800. Despite its success, the original windmill was still available as a backup. Today, the windmill still stands at this site, but the steam pump has long been removed.

Despite his failure in Rotterdam, Hoogendijk never gave up on steam-driven pumping. A second project was attempted. In this case, maintenance drainage for the 300-hectare (740-acre) Blijdorp Polder near Rotterdam was needed. Like John Hope's water garden, the Blijdorp Polder required the simpler constant lift pumping. After some negotiations regarding patent rights in the Netherlands, a Watt engine was selected. Construction began in 1786. The pump facility required the placement of 177 pine piles, driven to a depth of 15 meters (49 feet). The pump had a capacity of 0.8 cubic meters per second (28 cubic feet per second) and an efficiency of 1.7 percent. It was operational in September 1787 (van der Pols and Verbruggen 1996). This pump proved to be a success. In 1790 it was awarded a visit by Stadholder William V and his wife and two children—one of whom was the future King William I. Despite the success of this engine, it was used to drain the Blijdorp for only a short time. The local polder board was unwilling to take over its operation, primarily for political reasons.

In 1794 a Watt steam engine pump was selected for draining the Mijdrechtse Polder. The engineers planned to begin with a large pump diameter for initial low lift. As the polder was drained (and the lift increased), the pump would be replaced with one with a smaller diameter. Problems with insufficient coal supply, equipment failure, fire, dike seepage, an unfavorable political situation, and high operation costs caused the owners to abandon the entire project in 1812.

Figure 4-9: The Rotterdam Fire Engine building showing the beam arrangement after the two outer beams had been removed.
From Gemeente Archief Rotterdam.

Figure 4-10: Steam-driven pump in Heemstede.
From Gemeente Archief Rotterdam.

Figure 4-11: The operating principle of the Hellevoetsluis pumping station. The three main beams and three auxiliary ones can operate up to nine pumps.
Reprinted, by permission of Museum De Cruquius, from van der Pols and Verbruggen 1996.

In 1802 a steam pump was used for a dry dock at the Hellevoetsluis naval base. This application required a pump with varying lift capabilities. The anticipated lift would vary from 0 to 5.5 meters (0 to 18 feet). The solution was devised by hydraulic engineer Jan Blanken. Like Hoogendijk's Rotterdam engine of 25 years earlier, he used multiple pumps driven by a single steam engine, but in a mechanically much sounder way (van der Pols and Verbruggen 1996). The arrangement selected is shown schematically in **Figure 4-11**. The primary beam (connected directly to the engine) was attached to two other large beams by means of two pulleys with chains. When the primary beam moved upward, the two others moved downward. These three beams operated the three larger pumps. Three smaller pumps were connected at the midpoint of the main beams. Three more pumps were connected to secondary beams running off the main pump chain from the main beams. By connecting or disconnecting the chains attached to the pumps, the operator could pump water under a wide range of lift conditions. The steam engine and several pumps were manufactured by the Boulton and Watt Company in England. It is interesting to note that at this time England was at war with the Netherlands, yet they were able to provide technology for the Dutch naval fleet. The communications with Boulton and Watt never mentioned the intended use of the pump (Jan Verbruggen, personal communication). The pump was put into operation in 1802 and ran successfully for many decades.

In 1802 a Watt steam pump was used at a peat extraction site in the Krimpenerwaard Polder south of Gouda. Because of the poor peat quality of the area, this project (including the pump) was abandoned in 1813.

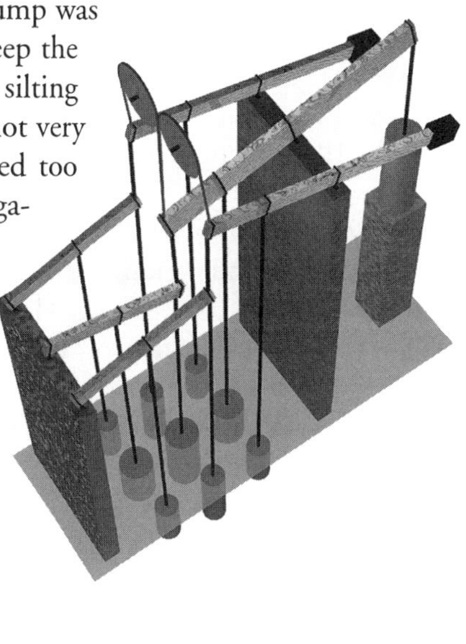

In 1807 another Watt engine was delivered to Katwijk to provide irrigation for lands dried up by the construction of a new outlet at the old mouth of the Rhine River. This pump was also to be used to keep the channel mouth from silting up. The pump was not very successful. It delivered too much flow for irrigation and eventually damaged the irrigation ditches. The operators of the sluices for the channel outlet were able to keep the channel silt-free without aid of the pump. The pump was sold in 1837, and a windmill was built for irrigation.

Steam-powered Archimedean screw pumps were installed to redrain the Wijde Wormer Polder north of Amsterdam. This polder was originally drained by windmills in the seventeenth century. In 1825 floods on the Zuiderzee turned the polder back into a lake. After the lake was drained again, the pumps were sold for other uses and windmills were used for maintenance draining.

In 1826 a pump was installed for draining water from the Zederik boezem to the Linge River near the Arkelse Dam. This was the first use of a steam-driven scoop wheel. Three Watt engines each drove a corresponding scoop wheel. Parts of this facility now reside at the Cruquius Museum (see "Places to Visit" at the end of this chapter).

In 1839 the Zuidplas Polder was reclaimed, using both wind- and steam-driven pumps. The pumps used were Archimedean screws driven off a horizontal steam engine. These pumps were effective in draining the polder, but the windmills were the primary source of maintenance drainage after the polder was dry.

Draining the Haarlemmermeer

At its largest the Haarlemmermeer covered 180 square kilometers (70 square miles) with an average depth of 4.5 meters (14.8 feet). Draining the Haarlemmermeer would require the removal of 800 million cubic meters (28.3 billion cubic feet) of water (van der Pols and Verbruggen 1996). This would never have been feasible without the use of steam power. But, considering steam power's many failures and only sporadic successes over a nearly 100-year period, one can understand why those wishing to drain the Haarlemmermeer were reluctant to commit to steam power. In fact, it is somewhat surprising that those undertaking the drainage of this lake chose to rely on steam power alone. The Zuidplas Polder, after all, was successfully drained with a combination of steam and wind power.

Over the years, there were significant objections to the various plans to drain the Haarlemmermeer. Even though the cities of Leiden and Amsterdam were threatened by flooding from the Haarlemmermeer, they were also concerned about other issues. Both cities regularly flushed their canal water into the lake to maintain the quality of their drinking water. Draining the lake could have a negative impact on the drinking water quality. The city of Leiden also held the fishing rights to the lake. The loss of the lake would thus be a significant economic loss to their community. In addition, the Rhineland regional water board was concerned about the loss of boezem storage volume. They relied on the Haarlemmermeer for both conveyance and storage of surface water at a regional level. In the end, the need for flood protection took priority, but, as was often the case, it took two floods for this to happen.

In November of 1836 a storm caused the Haarlemmermeer water to flood the land up to the outskirts of Amsterdam. Just one month later, another storm forced water into the streets of Leiden. Finally, the government took decisive action concerning the problem of the Haarlemmermeer. In 1837 King William I appointed a committee to make recommendations about solving this problem. The committee recommended a pumping arrangement that included a variety of types of pumps that used both steam and wind power. It included 79 windmills along with three 55-kilowatt (41,000-foot-lb-per-second or 74-horsepower) steam-powered Archimedean screws. The committee considered steam power for auxiliary pumping only. The King, who as a young man had visited the successful Blijdorp pump back in 1790, did not agree. He appointed a new committee to try again and to look more closely into the use of steam power. One of the members of that committee was G. M. Roentgen, who was the principal adviser to the King for industrial policy. He was also the managing director of the Etablissement Fijenoord engineering works, a company that manufactures steam pumps. It also included engineer D. Mentz, who could see no value in steam power. This committee became deadlocked over the type of pumps to use. Despite encouragement from the King, this committee was still unable to reach a decision. Finally, a third committee was appointed, which presented three alternatives and the corresponding estimated costs. The three alternatives were windmills alone, rotative

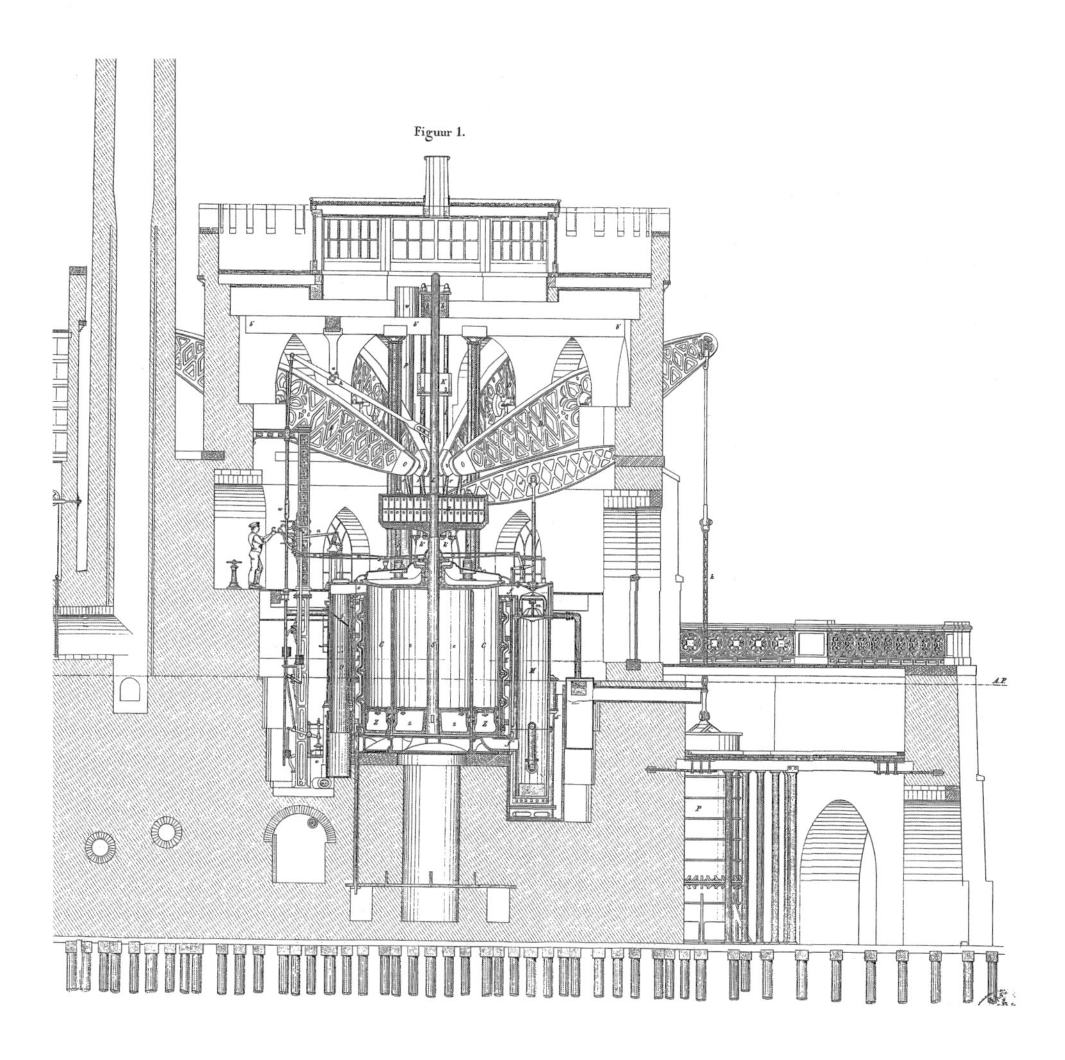

Figure 4-12: An 1852 cross section of the Cruquius steam pumping station. Note the single crosshead above the steam engine that drives pumps arranged in a circle around the engine. The boiler (not shown) is to the left.
From Gevers 1852.

steam engines driving scoop wheels or Archimedean screws, and three reciprocating Cornish steam engines driving piston pumps. The third alternative had the lowest construction, operation, and maintenance costs. The king accepted this alternative and reconstituted the committee to oversee the design and construction of the drainage facilities (van der Pols and Verbruggen 1996).

The first step in designing the project involved determining the required pump capacity. The choice of capacity was a compromise between the requirements of pumping the lake dry and the requirements of maintaining the lake in a dry state. A period of 14 months was selected for draining the lake. This meant that each of the three pumps needed a capacity of 22 cubic meters per second (780 cubic feet per second) at an average 2-meter (6.6-foot) lift. For maintenance draining, the capacity calculations were based on pumping out the greatest 30-day influx of rain and seepage within that same 30-day period. This resulted in a required maintenance capacity of 14 cubic meters per second (490 cubic feet per second) at a 4.5-meter (14.8-foot) lift (van der Pols and Verbruggen 1996). With this capacity the total number of pumping days per year was expected to be in the range of 50 to 60.

The number of pumping stations and the number of pumps in each station was determined by considering the range in discharge capacities and the need to always have backup units available. The final choice was to build three pumping stations, each powered by a single Inverted Sims Annular Compound engine rated at 260 kilowatts (190,000 foot pounds per second or 350 horsepower). Each of these engines was to drive up to 11 pumps. The 11 pumps were arranged in a circle around the engine and each pump was to have its own beam connected to a single crosshead. The crossheads weighed 650 kilonewtons (145,000 pounds) (Jan Verbruggen, personal communication).

The type of pump selected was a simple lift pump. Tests done on a wooden version of this pump connected to the Zuidplas Polder engines showed that this would work. **Figure 4-12** shows one of the Haarlemmermeer pumping stations, named "Cruquius" after the eighteenth century surveyor Nicolaas Cruquius. This image is part of a drawing made in 1852 (Gevers 1852).

To reduce the risk associated with this new design for steam-driven pumping, the planners decided to first build and thoroughly test one of the pumping stations before building the other two. The first facility was

▾ **Figure 4-13:** Miscast steam cylinder intended for Lake Haarlem pumping stations. Cylinder bore was 3.7 meters (12 feet). Photographed in the nineteenth century in Hayle, Cornwall, near the Harvey works where the Leeghwater and Cruquius engines were made.

▾ **Figure 4-14:** Drainage pattern in the Haarlemmermeer Polder. Three pumping stations— Lynden, Cruquius, and Leeghwater—were used to drain the water from the interior drainage ditches shown into the ring canal surrounding this former lake. Pumping station Halfweg was used to pump water from the ring canal into the IJ.

Redrawn from van de Ven 2004, Ch. 7, Fig. 12.

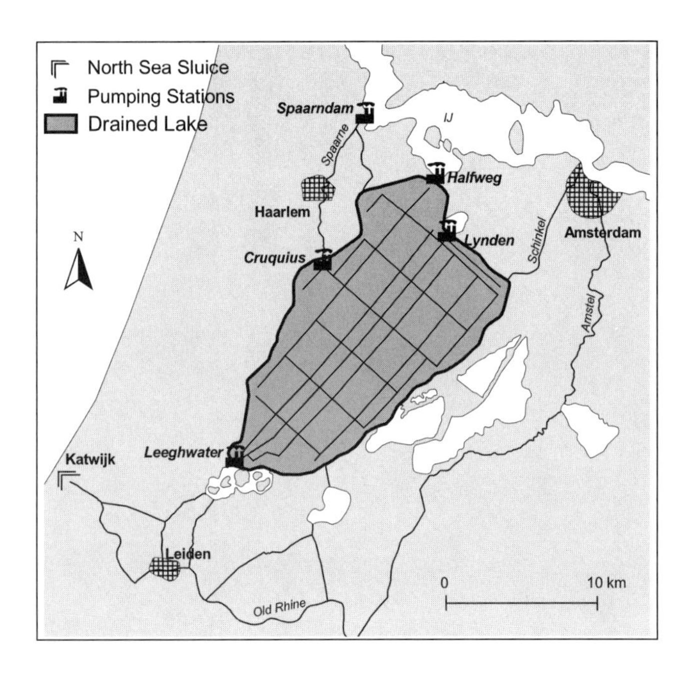

named after the famous millwright from the seventeenth century, J. A. Leeghwater. The Leeghwater steam cylinder had a bore of 3.66 meters (12.0 feet) and a stroke of 3.05 meters (10 feet) (see **Figure 4-13**). Its efficiency varied between 2.4 and 7.2 percent. It provided power for 11 pumps, each with a bore of 1.6 meters (5.25 feet) (van der Pols and Verbruggen 1996). Leeghwater took its first strokes in July 1845. In November of the same year, after some modifications, the working pumping station was demonstrated to King William II (William I abdicated in 1840). Tests of the pumping station showed that it could work, even with all 11 pumps operating under maximum lift conditions.

With Leeghwater working, the construction of the other two pumping stations began. They were named after two people who had, years earlier, proposed draining the Haarlemmermeer—Nicolaas Cruquius and F. G. van Lynden van Hemmen. The Cruquius and Lynden pumping stations were somewhat modified from the Leeghwater design. The primary difference was that the Cruquius and Lynden pumping stations used eight pumps instead of 11 but with larger bores [1.83 meters (6.00 feet)]. These two facilities were operational in the spring of 1848.

The 60-kilometer (37-mile) ring dike was constructed from the spoil material from digging the ring canal. The digging of the canal and construction of the dike began in 1840 and was completed in 1848. Pumping began as soon as the dike was complete. Pumping initially involved only the Leeghwater pumping station. Cruquius and Lynden were completed several months after the initiation of pumping. Because of various delays, it eventually took 39 months to drain the lake.

Figure 4-14 shows the configuration of what is now referred to as the Haarlemmermeer Polder. The main drainage canal runs along the axis of the polder, between the Leeghwater and Lynden pumping stations. Six secondary canals intersect the main canal, one of which is linked to the Cruquius pumping station. There are several routes for water to be discharged from the ring canal to the North Sea. Water can be discharged directly into the North Sea via the sluice gates at Katwijk. It can also be discharged into the IJ at either Spaarndam or Halfweg. The IJ is connected directly to the Zuiderzee which, in turn, is connected to the North Sea. The Spaarndam and Halfweg exits required the construction of additional steam pumping facilities.

▼ **Figure 4-15:** Cruquius pumping station. This is one of the three original Haarlemmermeer pumping stations. It has been converted into a drainage museum.

◢ **Figure 4-16:** Cruquius pump cylinder.

The Lynden and Leeghwater pumping stations were eventually modernized. In 1893 Lynden was converted and upgraded to two centrifugal pumps powered by modern steam engines. In 1912 Leeghwater was upgraded with a single centrifugal pump, powered by a diesel engine. These upgrades relegated Cruquius to standby status. After other modernizations, Cruquius was taken completely out of service in 1932. In 1973 the building was given "monument" status and eventually converted into a museum of drainage. In 1991 it was designated as an "International Historical Mechanical Engineering Landmark" by the American Society of Mechanical Engineers. The museum volunteers have recently restored the pumps for demonstrational purposes. The pumps are now driven by a hydraulic cylinder instead of the original steam engine. The photograph in **Figure 4-15** shows the Cruquius pumping station. **Figure 4-16** shows the top of one of the pump cylinders.

Prior to its draining, the Haarlemmermeer was a regional boezem. In other words, it was used for both storage and conveyance of surface water at a regional level. Water that was drained into the lake could then be discharged at low tides through sluice gates at Spaarndam, Halfweg, Katwijk, and Gouda. Draining the lake not only reduced the available boezem storage but it also increased the volume of water that needed to be handled by this boezem. As a result, additional pumps had to be placed at various locations to aid in discharging the region's excess water. The pump at Halfweg was a scoop wheel design driven by a horizontal steam engine. This facility has now been converted into a museum. Its scoop wheels and steam-driven engines, built in 1923, are still operational.

Settlements in the Drained Lake

The Haarlemmermeer ceased to exist as a lake in July 1852. Instead, there were 180 square kilometers (70 square miles) of land with soil composed mostly of blue marine clay. Adequate drainage required that a significant proportion of the land be allocated for conveyance (i.e., ditches and canals). In the recently reclaimed Zuidplas Polder, the ratio of drainage conveyance area to dry land was 1:10. In an effort to regain as much of their investment as possible, the Dutch government decided on a parcelization scheme that resulted in a conveyance/dry land ratio of 1:32 (van de Ven 2004). This provided more revenue, but it resulted in worse drainage. An additional problem associated with drainage was that the water levels in the ditches and canals were nearly constant throughout the polder. Since the water was drained out of the polder at the perimeter, the water levels in the ditches and canals had to be somewhat higher at the center of the polder where the ground surface was the lowest. As a result, drainage was very poor in the polder, especially near the center. Diseases associated with the wet conditions—cholera, typhus, and malaria—were a significant problem for many years. Infant mortality rates in 1861 were reported to be double that outside the polder.

Other problems existed as well. Before the land could be sold, it was left fallow, allowing weeds to grow and take root. The combination of wet, stiff clay and existing weed growth made initial cultivation very labor intensive. The roads in the polder were inadequate, and the canals were too shallow to carry agricultural products out by boat.

In August 1853 the first parcels were sold. Initially, 86 percent of the land was purchased by speculators and large landowners. The first residents, as a rule, leased the land. In 1856, with a population of about 1,800, the Haarlemmermeer Polder appointed its first governing board. Also in 1856, gravel was first used to improve the roads in the polder. By 1857 the population increased to about 5,200. The residents were eventually able to establish their own polder board within the regional Rhineland water board. With this administrative structure in place, improvements could finally be made. It took almost 25 years for most of the problems to be worked out and the region to become prosperous.

The Haarlemmermeer Today

Today, the Haarlemmermeer Polder has three distinct uses. Agriculture still dominates a large portion of the polder. The central town of Hoofddorp grew from a small farming village to an important location for international business. The most significant feature of the Haarlemmermeer Polder today is the presence of Schiphol Airport, one of the world's busiest. Located where navy battles once took place, it occupies nearly one-quarter of the former lake.

Places to Visit

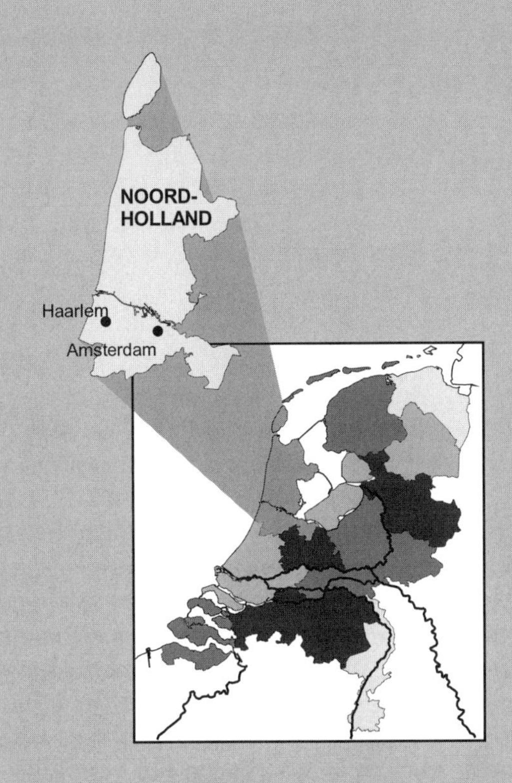

Cruquius Museum

DESCRIPTION: This museum is located in one of the three original pumping stations on the Haarlemmermeer. There are displays about steam pumping as well as land reclamation and water management. You can view the pump mechanism in operation.

LOCATION: Halfway between Hoofddorp and Heemstede along N201 in the province of Noord-Holland.

INTERNET: (1) www.museumdecruquius.nl: This is the official museum Web site with information in English. (2) www.cruquiusmuseum.nl: This Web site provides information about the design and history of the steam pump, including a complete version in English.

EMAIL ADDRESS: info@MuseumDeCruquius.nl.

ALONG EXCURSION: Haarlemmermeer.

Steam Pumping Plant Halfweg

DESCRIPTION: This museum is located at the pumping station that lifted water from the Haarlemmermeer ring canal into the IJ. Built in 1853, it is the oldest operational steam pumping station in the world. It is put into operation for demonstrations approximately 12 days per year. The calendar of pumping days is provided on the Web site.

LOCATION: Halfway between Amsterdam and Haarlem along N200 in the province of Noord-Holland.

INTERNET: www.stoomgemaalhalfweg.nl (complete English version).

EMAIL ADDRESS: info@Stoomgemaalhalfweg.nl.

ALONG EXCURSION: Haarlemmermeer.

Historical Museum Haarlemmermeer

DESCRIPTION: Museum focusing on the development of the Haarlemmermeer Polder area. It covers some of the technological aspects but focuses more on the social development of the area. The museum includes some display descriptions in English. Be sure to get a copy of the driving directions from the "Routebeschrijving" page on the Web site.

PHONE: 0-23-5620437 (from the United States 011-31-23-5620437).

ADDRESS: Kruisweg 1403, 2131 MD Hoofddorp (province of Noord-Holland).

HOURS: April through September: Tuesday through Sunday, 1:00 to 5:00 p.m.; October through March: Saturday and Sunday, 1:00 to 5:00 p.m.

ENTRY FEE: € 3.00 (free with Dutch Museum Card).

INTERNET: www.historisch-museum-haarlemmermeer.nl (in Dutch only).

ALONG EXCURSION: Haarlemmermeer.

5 Reclamation of the Zuiderzee

The growth of the Zuiderzee from an inland lake to a large sea arm has been mentioned several times. This chapter focuses on the details of its reclamation and subsequent development. The reclamation of the Zuiderzee took place in the twentieth century. By that time, the tools and machines used had advanced beyond wind and steam power. This allowed for the implementation of projects on a much larger scale.

Figure 5-1 shows the history of the growth of the Zuiderzee. The top-left panel shows Lake Flevo in the Roman era, prior to any significant man-made changes in the region. Lake Flevo was fed by several streams, the most significant being the IJssel River—one of the branches of the Rhine. At this time, the level of the lake was above that of the North Sea. The top-right panel shows the configuration after the North Sea's influence opened up the lake, creating a tidal sea arm. Around this time it became known as the Zuiderzee (or Zuyderzee), literally translated "South Sea." This growth was largely due to an erosion of the peaty shoreline and a rise in sea level. The progression from 1300 to 1850 shows expansion of the Haarlemmermeer (the lake southwest of Amsterdam), growth of the Zuiderzee, but elimination of the lakes in Noord-Holland. Today a significant portion of the Zuiderzee has returned to dry land.

Historical Development of the Zuiderzee Reclamation Plan

Like any large-scale project, the reclamation of the Zuiderzee took many years from first inception to completion. The first plans were published as early as the seventeenth century. Encouraged by the success of lake drainage earlier in the century, Hendric Stevin published plans in 1667 for draining the Zuiderzee. At the time, the Zuiderzee was composed of all of the water behind the chain of Frisian islands, including the part that is now called the Waddenzee. Progressing from draining/reclaiming Lake Beemster (the largest of the Noord-Holland lake reclamations discussed in Chapter 3) at 72.2 square kilometers (27.9 square miles) to draining the Zuiderzee at 6,700 square kilometers (2,600 square miles)—nearly 100 times larger—would be possible only after several centuries of technological advancement.

In the nineteenth century the number of proposals for the reclamation of the Zuiderzee increased. **Figure 5-2** shows four of these plans (van der Wal 1922). Kloppenburg and Faddegon proposed damming the Zuiderzee at the narrowest point and draining everything that lay on the southeast side of the dam. Their plan included reclamation of the IJ (Zuiderzee arm north of Amsterdam) as well as the creation of a peripheral canal to drain the rivers that discharge into the Zuiderzee. Van Diggelen's plan proposed reclaiming nearly all of the Zuiderzee and leaving a peripheral lake for river drainage and water storage. Leemans addressed only the southern part and included a network of open storage areas. Opperdoes Alewijn and Kooy suggested reclaiming only the parts of the Zuiderzee near the shore. By 1877 when Leemans's plan was proposed, the Nordzee Kanaal and the IJ reclamation were already completed.

▾ **Figure 5-1:** Growth of the Zuiderzee from the Roman era to the present.

Redrawn from SWAVN 1984, Figs. 5 and 6, and van Duin and de Kaste 1990, Fig. 7.

◢ **Figure 5-2:** Several plans for the reclamation of the Zuiderzee.

Redrawn from van der Wal 1922, Fig. 2, and van de Ven 2004, Ch. 10, Fig. 3.

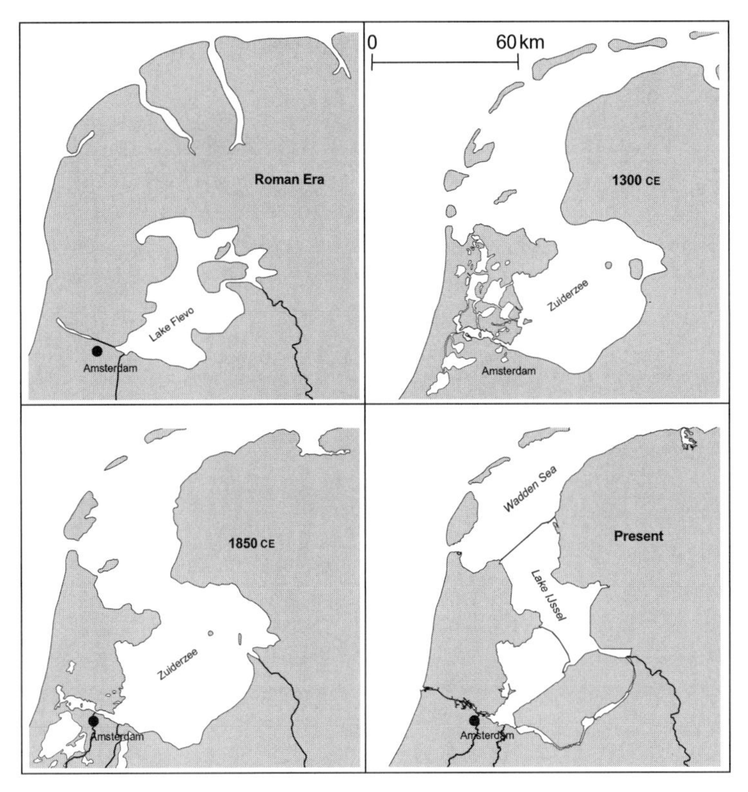

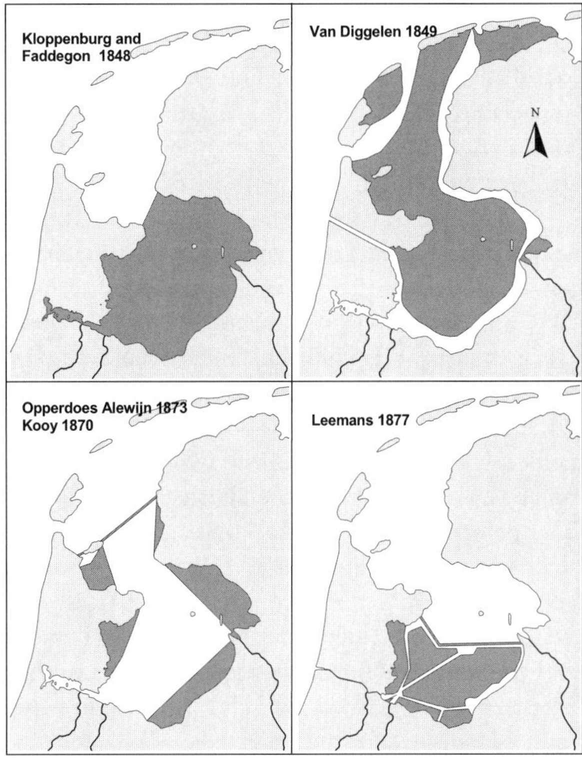

All of the plans presented by the mid 1880s lacked good scientific and engineering data about the nature of the Zuiderzee. Furthermore, the central government made no effort to pursue this project. As a result, a group of prominent citizens formed an association in order to find the funds to pay for the research and design of the Zuiderzee reclamation. The association—known as the Zuiderzee Association—was formed in 1886. Their first task was to hire an engineer to lead the project.

Cornelis Lely (1854–1926) was a graduate of the Delft University of Engineering. The first decade of his career was largely unsuccessful, as he was unable even to attain the level of "assistant engineer." As a result he had difficulty keeping food on the table (van Veen 1955). Despite these struggles, Lely developed a strong interest in the reclamation of the Zuiderzee. Soon after the Zuiderzee Association was formed, he responded to an employment advertisement placed by the Association and was eventually hired to lead the project.

In 1891 Lely published the first technically feasible master plan for the reclamation of the Zuiderzee. **Figure 5-3** illustrates the Lely plan (van der Wal 1922). It involved the construction of a dam (labeled Afsluitdijk) at the point where the Zuiderzee turned inland. Thus the area behind the Frisian Islands would remain part of the sea (later to be renamed the Waddenzee), and the dam would close off the rest of the Zuiderzee from the influence of the North Sea. This dam would be equipped with outlet sluices to allow water to discharge toward the North Sea. This structure provided a number of benefits. First, it better protected the areas surrounding the Zuiderzee from flooding. Second, it provided for better land-transportation routes between the Noord-Holland and Friesland provinces. Third, with freshwater entering

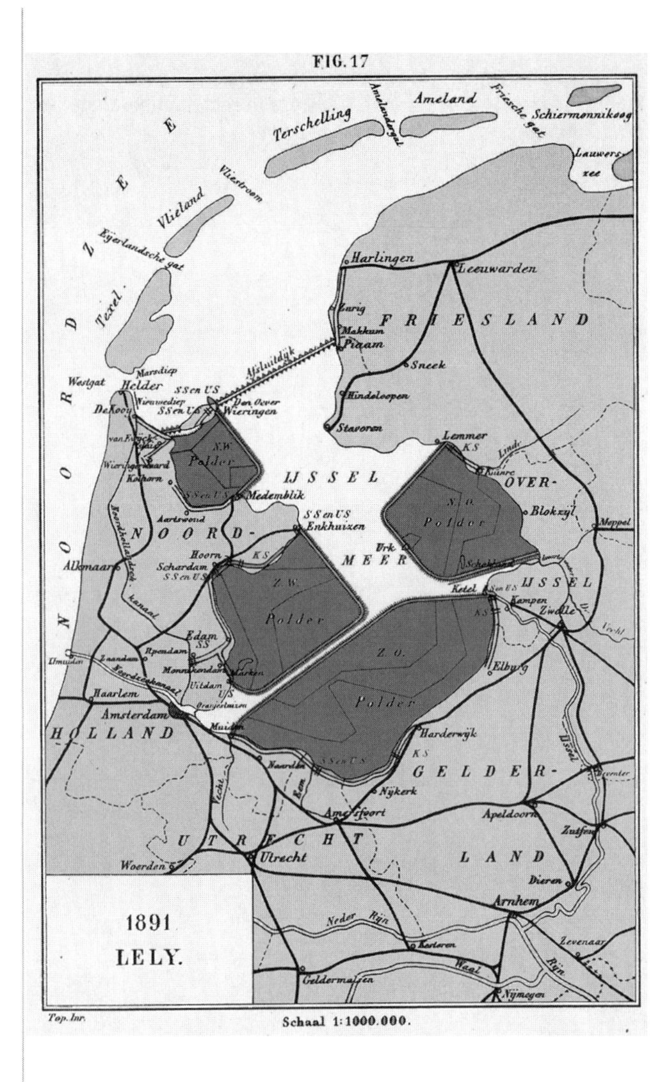

Figure 5-3: Lely Plan for the reclamation of the Zuiderzee.
From van der Wal 1922, in the collection of the Sociaal Historisch Centrum voor Flevoland te Lelystad.

from the IJssel River, the dam would turn the saltwater Zuiderzee into a freshwater lake. This lake was eventually called the IJsselmeer. Although this had negative impacts on the region's fishing industry, it greatly benefited agriculture in the region, which had previously struggled with salinization problems.

After the Zuiderzee was dammed, four large polders were to be built. They would be constructed by building a dike from the mainland at one location, into the lake, and back to the mainland at a second location. Once the dike was complete, water would be pumped out of the endiked area and into the newly created freshwater lake. Lely's plan for the polders had two goals. The polders had to be located in areas with clayey soils, as this soil was good for agriculture and had limiting seepage. The polders also had to be located in areas that were shallow to reduce the volume of water to be pumped. Fortunately, both criteria could be met. **Figures 5-4** and **5-5** show the soils and depths in the Zuiderzee (van Duin and de Kaste 1990) along with the outlines of the four proposed polders. Note that Lely's plan also incorporates four islands—Marken, Wieringen, Urk, and Schokland—to the mainland. Lely's plan included a construction timeline of 32 years.

The freshwater lake formed behind the barrier dam was essential to the plan's success. The level of the North Sea varied because of tides and storms. This meant that a significant amount of storage was needed for the natural rivers that continued to flow into the newly formed lake. When the sea level is high, the outlet sluices are closed and the lake level begins to rise slowly as the rivers discharge into it. Once the sea level drops below that of the lake, the discharge sluices can once again be opened and the lake water is allowed to discharge. Lely calculated that a lake of 800 square kilometers (310 square miles) would provide sufficient water storage. This estimate was later increased to 1,200 square kilometers (460 square miles) (van de Ven 2004).

Despite the excellence of Lely's plan, it still took several decades for implementation to begin. Many issues had to be resolved. Some of these issues were technical in nature; others were related to the impact that the plan would have on those living in the region, such as the loss of fisheries. The project also required considerable political support. During the end of the nineteenth century and the beginning of the twentieth, three events occurred that gave the project the momentum it needed.

First, the political situation swung in favor of the draining of the Zuiderzee. In 1891, after Lely finished his work with the Zuiderzee Association, he was appointed to the cabinet position of Minister of Water Management, Trade, and Industry. In 1914 he was in his third term in this position (now under the new title of Minister of Water Management). He was now in a position to reintroduce legislation to begin the construction of the Zuiderzee reclamation. Second, the Dutch began to realize that they needed more farmland to feed the growing population, which draining part of the Zuiderzee would provide. The Dutch population realized this firsthand when food shortages occurred during World War I. Third, the region was struck by a flood. The region had been free of significant flooding for many years. The problems of the past were too soon forgotten. In 1916 a storm surge raised water levels on the Zuiderzee and threatened the safety of those living in Amsterdam. The act to begin the reclamation was unanimously approved

▾ **Figure 5-4:** Soils at the bottom of the Zuiderzee. The four proposed polders are also shown.
Redrawn from van Duin and de Kaste 1990, Fig. 14.

◢ **Figure 5-5:** Depth of the Zuiderzee. The four proposed polders are also shown.
Redrawn from van Duin and de Kaste 1990, Fig. 15.

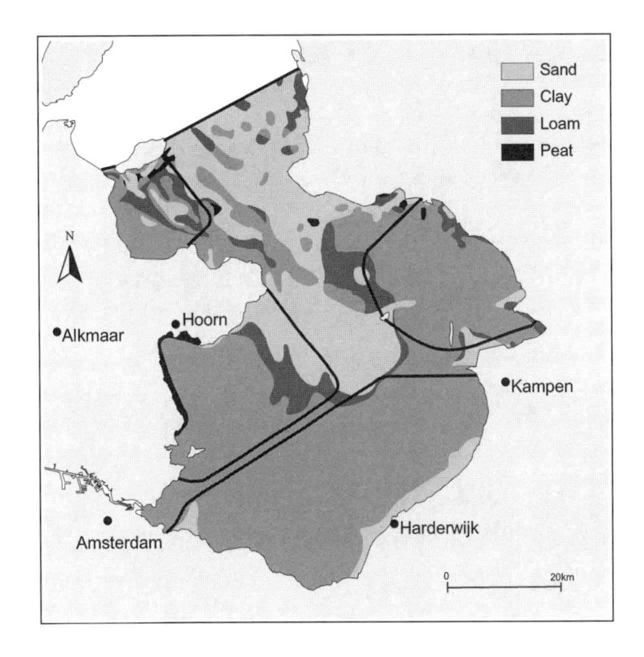

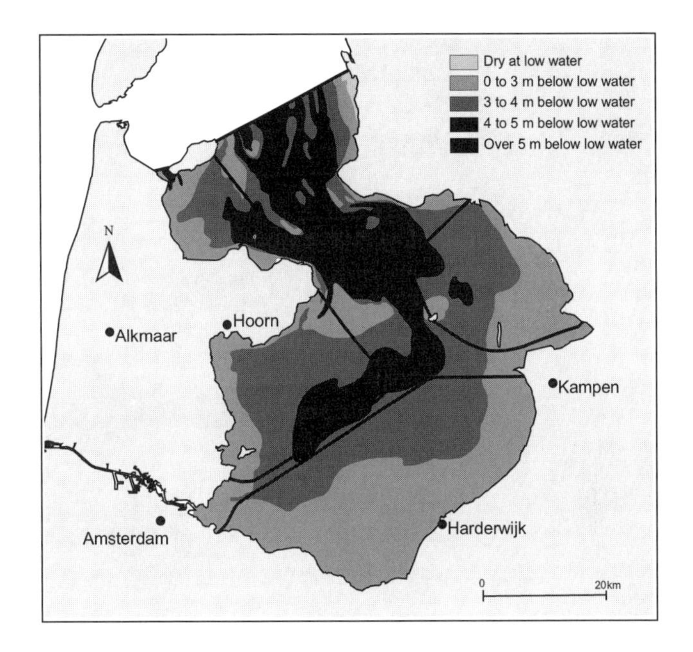

by legislature in 1918. Lely decided that the best way to proceed was to create a special independent authority for doing further engineering study and design. Construction on the first phase of the Zuiderzee reclamation was finally scheduled to begin in 1920.

Sequence of Activities

The initial timetable for construction of all of the elements of Lely's plan was never realized. War and an economic recession caused delays in the plans. Also, parts of the plan were never completed. Specifically, the Markerwaard Polder (Z.W. Polder in **Figure 5-3**) was started several times but never finished. The basic approach to sequencing the construction activities started with creating the barrier that closed off much of the Zuiderzee from the North Sea, thus creating a freshwater lake and providing additional security to the region. Once the region was protected from the North Sea, the polders could be constructed. The construction sequencing approach started with the smallest polders and gradually worked toward the largest polders. This meant that lessons learned early could be applied to larger polders later. This also delayed the financial investments as long as possible.

The overall sequence of construction is given in **Table 5-1**. The names given in this table can all be found in the map shown in **Figure 5-6**. Details on the major parts of the construction and design will follow in later sections.

Construction of the Barrier Dam

The goal of the Barrier Dam (or "Afsluitdijk") was to block off the southern half of the Zuiderzee from the influence of the North Sea. The first step, a dike from the northern tip of the Noord-Holland mainland to the island Wieringen, passed through a tidal channel called the Amsteldiep. Next, the 30-kilometer (18.6-mile) Barrier Dam from Wieringen to the Frisian coast would be built (see map in **Figure 5-6**). This dike passed through two deep tidal channels. On June 29, 1920, construction

Table 5-1: Zuiderzee Reclamation Timeline.

Started	Completed	Activity
1920	1925	Construction of Amsteldiep Dike, connecting the island of Wieringen to the mainland.
1927	1932	Construction of the Barrier Dam between the island of Wieringen and the Frisian coast. This turns the southern half of the Zuiderzee into a freshwater lake called the IJsselmeer. The northern half of the Zuiderzee is eventually renamed the Waddenzee.
1927	1927	Construction of the research polder Andijk.
1927	1930	Construction of the Wieringermeer Polder. Construction sequencing was based on polder size. Since the Wieringermeer was the smallest, it was built first.
1937	1942	Construction of Noordoost Polder. This became the second smallest polder after the original 1918 plans were revised in 1925.
1941	1941	The dike construction for the Markerwaard was begun. Construction started at the island of Marken and proceeded northwest. The occupying German authorities stopped construction.
1950	1957	Construction of Oostelijk Flevoland. Another revision of the plans, made after World War II, split the Flevoland Polder into two separate pieces. The Flevoland pieces were now smaller than Markerwaard, so construction began there. Oostelijk Flevoland construction funding was supplemented with Marshal Plan money.
1956	1959	Markerwaard construction continued. This construction included a dike connecting Marken to the mainland, pits for lock complexes at Enkhuizen and Lelystad, and the dike running in the direction from Amsterdam to Lelystad. Construction was stopped on Markerwaard as the need for additional housing increased, raising the priority of completion of Zuidelijk Flevoland.
1959	1968	Construction of Zuidelijk Flevoland. The dike constructed for the east side of Markerwaard now became the west edge of Zuidelijk Flevoland.
1963	1975	Work started again on the Markerwaard with the construction of the dike from Enkhuizen to Lelystad. This has provided an important east-west transportation route for the region. No further work has been done on the completion of Markerwaard. In 1991 the government ended discussion of the Markerwaard.

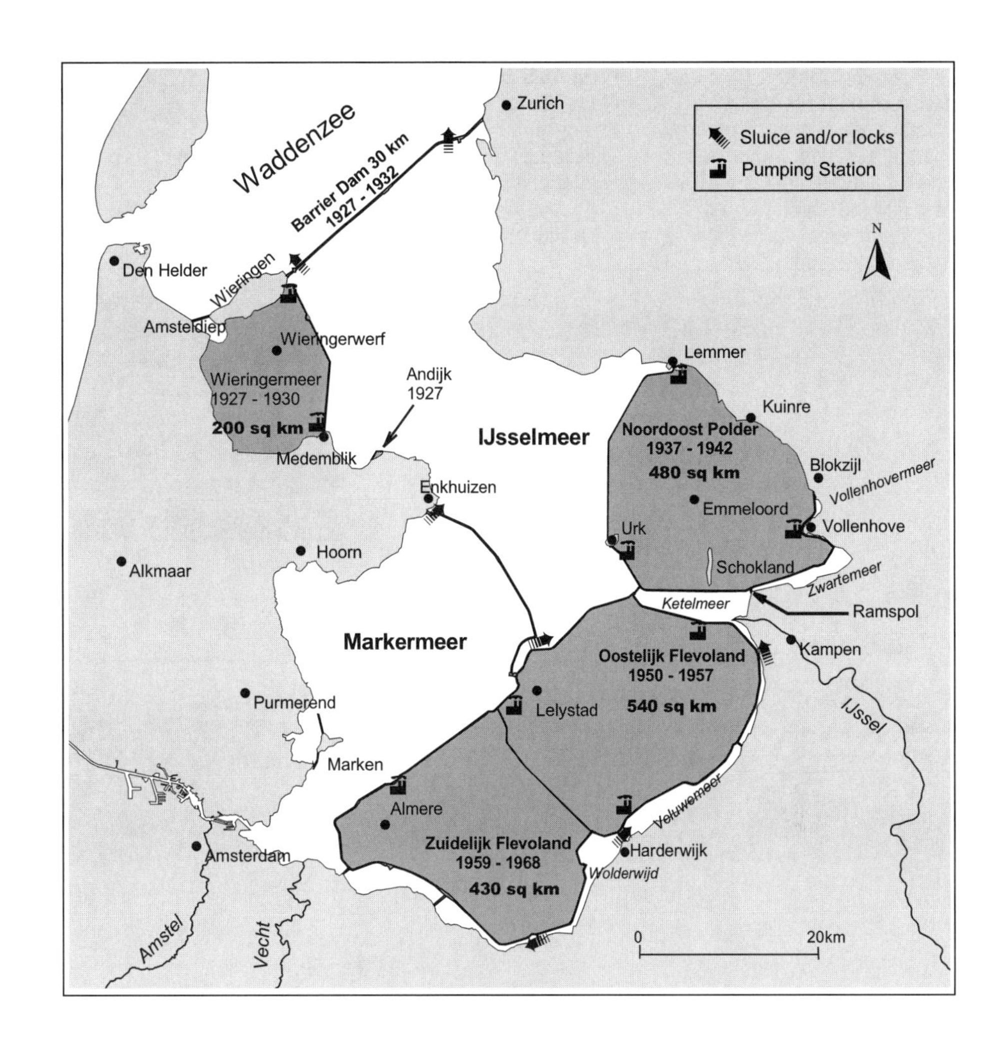

Figure 5-6: Layout of the IJsselmeer polders.
Redrawn from IDG 1994, p. 17.

began on the 2.5-kilometer (1.55-mile) Amsteldiep Dike. This was the first step in a very large project that would eventually take decades to complete.

Prior to any construction, a great deal of engineering data had to be collected. The sea inlets were sounded several times. This provided information about the effect that the dam would have on currents. Some soil borings had already been taken as early as 1877 (van de Ven 2004). Many more borings were taken to determine the soil types as well as the bearing capacity of these soils. With this information in hand, the engineers designing this structure made calculations to determine the discharge capacity needed for the outlet sluices. Scale models of the sluices were constructed to better understand the impact that the structure would have on currents and scouring. The initial physical models were built at the hydraulics laboratory of the Technical University of Karlsruhe, Germany. After 1927 the models were moved to the Hydraulics Laboratory at the Technical University in Delft.

A cross section of the Barrier Dam is shown in **Figure 5-7**. The shape and height were designed for the typical storms that occurred in the period from 1825 to 1926. The slope of the face of the dike was set at 4:1, based on model testing. The core of the dike was constructed of boulder clay, which was found in abundance in the Zuiderzee area. This clay, deposited in large lumps, can withstand water velocities of up to 4 meters per second (13 feet per second) (van de Ven 2004). **Figure 5-8** is a photograph taken in 1932 showing the boulder clay being placed at one of the closure points. The larger part of the dike body cross section consists of sand removed from the area using dredgers and transported in barges. To support this structure, the foundation of the dike needed proper preparation. The foundation soils consisted of 4 to 15 meters (13 to 49 feet) of clay. In some locations the foundation clays were too weak to support the structure. They had to be removed and replaced with sand.

The submerged part of the dam needed to be protected from scouring. This was done by sinking a mattress constructed of osier fascines onto the dike body (as shown in **Figure 5-7**). Fascines are long bundles of wood

▾ **Figure 5-7:** Barrier Dam cross section. The three water levels shown on the Waddenzee side are the low tide [–1.0 meters (–3.3 feet)], high tide [+0.6 meters (+2.0 feet)], and storm surge [+3.5 meters (+11.5 feet)].

Redrawn from van de Ven 2004, Ch. 10, Fig. 6.

▸ **Figure 5-8:** Closing the gap of the Vlieter, May 27, 1932. Photographer M.L.D. Vliegkamp "De Kooij," Fototechnische Dienst.

From the collection of Sociaal Historisch Centrum voor Flevoland te Lelystad.

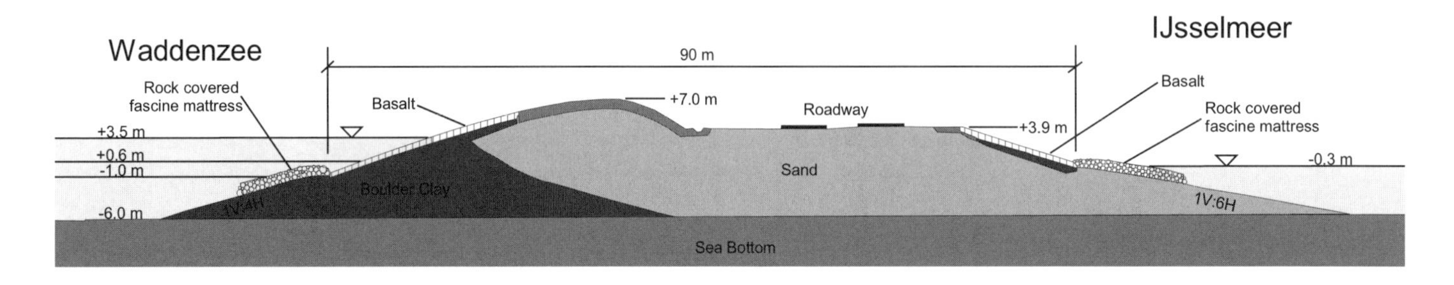

◂ **Figure 5-9:** Fascine mattress used to protect the Barrier Dam.

From Rijksarchief in Flevoland, photo archive no. RYP-1994.72.

▾ **Figure 5-10:** Installation of Basalt blocks.

From Rijksarchief in Flevoland, photo archive no. RYP-1994.163.

Figure 5-11: The barrier dam today.

bound together. The common type of wood used was osier—a type of willow with very pliable branches. After the mattress was constructed, it was floated into place and sunk using stones. The photograph in **Figure 5-9** (taken in 1930) shows one of these fascine mattresses being prepared for sinking. To protect the exposed face of the dike against the erosive action of waves, part of it was lined with basalt blocks. Constructing this basalt layer was a very labor-intensive process, as can be seen in the photograph in **Figure 5-10**.

The alignment of the longer portion of the Barrier Dam can be seen in **Figure 5-6**. The location of the north end of the dam, at Zurich, was chosen because of the quality of soil found there. The two bends near the north end allow the dike to cross one of the tidal channels in a direction perpendicular to the channel. One of the two sluice complexes is also located there.

Water from the IJsselmeer had to be discharged through the Barrier Dam into the Waddenzee. For this reason, two outlet sluice facilities were constructed. These facilities also included locks for boats to travel between the IJsselmeer and the Waddenzee. Each sluice gate is 12 meters (39 feet) wide. The Stevin sluice located near the southern end has 15 sluice gates, and the Lorentz sluice at the north end has ten gates. These allow for a discharge capacity of up to 5,000 cubic meters per second (177,000 cubic feet per second) (van Duin and de Kaste 1990). These outlet facilities were constructed in their own temporary construction pit. This required the construction of a dike surrounding the construction site. The water was then pumped out from within the dike to provide dry ground for construction.

Work began on the Amsteldiep Dike in 1920. Because of the fading of the memory of previous floods and the economic crisis of the 1920s, work was performed at a slower pace starting in 1922. The Amsteldiep was finally closed to flows in 1924, and work was completed in 1925. Work began on the 30-kilometer (18.6-mile) Barrier Dam in 1927. This was completed on May 28, 1932. **Figure 5-11** shows a recent photo of the Barrier Dam. It was originally planned to include a two-lane road and railroad line. The railroad was never built. Instead, a divided highway was built on the dam.

Polder Construction and Land Conditioning

The first step in the construction of each polder was the construction of the surrounding polder dike. **Figure 5-12** shows a cross section of the western section of the Noordoost Polder dike. Since the polders were protected from the tides and storms of the North Sea by the Barrier Dam, the polder dikes did not have to be built as high. Construction of these dikes was much easier than that of the Barrier Dam because they no longer had to contend with tidal flows. First, unstable soils were removed. Next, a layer of sand fill was placed. Then the abutments of the dike—two parallel strips of boulder clay—were laid down, a large one on the IJsselmeer side and a smaller one on the polder side. The space between the two clay foundations was then filled with sand, finishing the dike body. Riprap, asphalt, and basalt revetments were then placed as needed. The photograph in **Figure 5-13** shows the construction of one of these dikes. In the photograph, fascine mattresses sunk with stone are being placed on what will eventually be the polder side of the dike.

Pumping facilities were constructed next. These were located at the discharge points for the planned primary drainage canals. Early in the design of the pumping facilities, the engineers needed to compute necessary pumping capacity, in particular, the capacity to maintain the water levels once the polder was dry. Detailed calculations were made regarding the amount of water to be removed, including precipitation, seepage, and evaporation. Finally, the calculations accounted for the available storage in the canals and the permissible water surface rise. The engineers planned to keep the water level of the Wieringermeer Polder at 1.4 meters (4.6 feet) below dry ground. The maximum allowed rise during periods

▸ **Figure 5-13:** Construction of an IJsselmeer Polder dike.
From Rijksarchief in Flevoland, photo archive no. RYP-1994.345

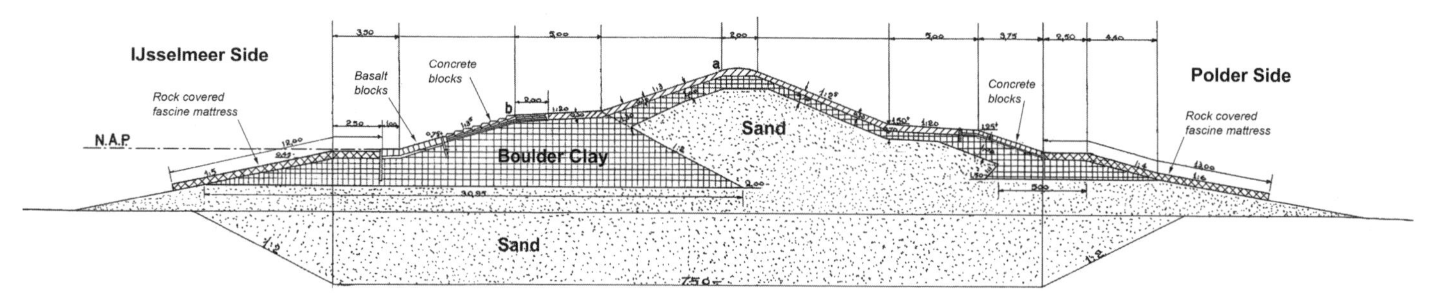

▾ **Figure 5-12:** Cross section of a Noordoost Polder dike.
From Rijksarchief in Flevoland.

without pumping was 0.2 meter (0.6 foot). The calculations were based on 8 millimeters per day (0.3 inch per day) of precipitation, 1 to 2 millimeters per day (0.04 to 0.08 inch per day) of groundwater seepage, and a precipitation uncertainty of 3 to 4 millimeters per day (0.12 to 0.16 inch per day) (van de Ven 2004). The construction of the pumping station near Medemblik in the Wieringermeer Polder, named after Cornelis Lely, is shown in **Figure 5-14**.

While the polders were still underwater, the main drainage canals were dredged. These canals can be seen in **Figure 5-15**, which shows the Wieringermeer Polder near Medemblik. Here you see the town of Medemblik, the newly dry polder, and the "Lely" pumping station (also shown in **Figure 5-14**). As shown in this photo, the canals leading to the pumps were already in place.

A number of problems occurred during the drainage of the Wieringermeer Polder that needed to be corrected as additional polders were constructed. Drainage ditches were dug in the polder before it was completely dry. When the polder was only partly drained, erosion occurred at the top of the ditch bank by water flowing into the ditch from the nearly drained fields. Furthermore, small dams, created by the ditch spoil piles, would suddenly burst, causing the ditches to erode even more. Once the polders were declared dry, the process of curing the soil started.

Reclamation of the Zuiderzee presented some challenges not faced in previous lake draining projects. In particular, soils that had been saturated by salt water for centuries were to be converted for agricultural use. Small areas along the coast had been reclaimed from saltwater inundation, but the Zuiderzee project was on a much larger scale. Based on these smaller coastal reclamations and on other studies, procedures were developed for draining and curing Zuiderzee bottom lands. But before these methods could be implemented, they needed to be tested and refined using a full-scale experiment. For this reason, a small, 40-hectare (100-acre) research polder near the village of Andijk was constructed in the Zuiderzee in 1927. It was located just north and west of the town of Enkhuizen (see **Figure 5-6**). The Andijk research polder was completed only four years prior to the first major polder—Wieringermeer. It provided some good information about the potential success of curing land for agricultural use. Additional research continued in the larger polders.

The first step in preparing soil for agricultural use was to remove excess water, lowering the water level to that of the former sea bottom. For the land to be usable, the water had to be lowered an additional 1.0 to 1.5 meters (3.3 to 4.9 feet). This was accomplished partially through artificial means—the use of drainage ditches or tiles—but the water was also removed by evaporation

▾ **Figure 5-14:** Construction pit for the pumping station "Lely" near the town of Medemblik.

Photographer M. L. D. Vliegkamp "De Kooij," Fototechnische Dienst. From the collection of Sociaal Historisch Centrum voor Flevoland te Lelystad.

▾ **Figure 5-15:** View of the nearly dry Wieringermeer Polder. The "Lely" pumping station shown in Figure 5-14 is visible, near the center of the photo, where three drainage canals intersect at the polder dike. The city of Medemblik is in the foreground.

From AVIODROM Aerial Photography, Lelystad, NL.

near the surface of the soil and by transpiration through the vegetation that was planted soon after the surface water was drained.

The first plants were reeds, sown from airplanes. The reeds grew well in the unmatured soil. They extracted a great deal of water from the soil, and their roots improved the bearing capacity and brought organic material into the soil.

When the soil was dry enough for equipment to work on the surface, drainage ditches were dug. After the reeds were removed, the first crop grown was rape—an herb of the mustard family grown as a forage crop and for its seeds, which yield rapeseed oil or canola oil. Rape grew well in wet soils with high nitrogen content. It suppressed the reeds and could be sown in the summer when the land was accessible. The next crop planted was usually a cereal such as wheat, barley, or oats. Crops such as flax and pulses (peas, beans, and lentils) followed them. At this time the ditches were replaced with drainage tiles. Finally, the land was ready to be handed over to the farmers.

On average it took about five years before any newly drained land was cured, making it suitable for normal agricultural practices. Development within the polder was done in stages. Agricultural land was placed in use at a rate of 3,000 to 4,000 hectares (7,400 to 9,900 acres) per year (van Duin and de Kaste 1990). This meant that it took ten to twenty years for all of the agricultural land to come into agricultural use in the larger polders.

It soon became evident that subsidence due to clay shrinkage and consolidation was a problem. In some areas the land dropped by as much as 0.7 meter (2.3 feet) within the first few years of reclamation. As a result, each area was assigned two groundwater levels—one for immediately after the polder was drained and a second for a later period, after several years of consolidation. All of the facilities that controlled the water levels, including the main pumping stations, had to be designed for these water level variations.

Construction of the Wieringermeer Polder

Wieringermeer was the first large area reclaimed (see **Figure 5-6**). Construction began in 1927. As can be seen in the map, this was the easiest of the polders to build. Old land bordered the polder on three sides. The dike had only to cover the distance from the tip of the Wieringen Island to the mainland near Medemblik. The dikes were completed by 1929. The polder was dry by August 11,

1930, after pumping 700 million cubic meters (25 billion cubic feet) of water, thereby creating 200 square kilometers (77 square miles) of new land (see photo in **Figure 5-15**). Two pumping stations kept this polder dry. Pumping station Leemans at Den Oever is diesel powered while pumping station Lely (shown under construction in **Figure 5-14**) was electrically powered. They had a combined capacity of 28.3 cubic meters per second (1,000 cubic feet per second) (van Duin and de Kaste 1990). The polder sloped from northwest to southeast. As a result, there were four separately controlled water levels within the polder. The greatest elevation difference between the canal water levels and the outside IJsselmeer water was 6 meters (20 feet).

A unique feature of this polder was that a refuge mound was constructed near the center in case one of the dikes was breached, causing the polder to flood. Ironically, this is the only one of the IJsselmeer polders ever to be completely flooded. In 1945 the retreating German army bombed the dike, filling the polder with water.

The polder's primary purpose was agriculture. The standard plot was 20 hectares (49 acres). These were generally 250 meters by 800 meters (820 feet by 2,600 feet). The front of the plot was accessible by road, while a canal defined the back of the plot. This canal was large enough for boats to transport agricultural products. The sides of most of the plots were drainage ditches. Individual farms consisted of one to three plots, depending on the nature of the farm. The ownership of the land remained in the hands of the government and the land was leased to the farmers.

Most of the design effort in the planning of the Wieringermeer Polder was concentrated in hydraulics (dikes, ditches, pumps, etc.) and land preparation. The planners assumed that the towns necessary for supporting agricultural activities would "sprout" spontaneously at important road crossings. When it was clear that this was not going to happen, the government realized that the construction of towns was going to be an important part of developing the new polders. Three towns were initially placed close to the center of the polder—Middenmeer, Slootdorp, and Wieringerwerf. These turned out to be too closely spaced. Their service areas overlapped. Some farmers had only a short distance to travel while others had too large a distance to travel. As a result of this, a fourth town, Kreileroord, had to be built later (Meijer 1992). (Note: Wieringerwerf is the only one of these towns shown on the map in **Figure 5-6**.)

Construction of the Noordoost Polder

The second large polder to be created in the former Zuiderzee was the Noordoost (Northeast) Polder (see **Figure 5-6**). The dike runs from the town of Lemmer on the North to Blokzijl on the east. The central town is Emmeloord, which is surrounded by ten smaller villages. The polder includes two former islands, Schokland and Urk. Two rivers in the old land needed to be dammed and diverted. Two peripheral lakes were formed at the east end of the dike, the Zwartemeer and the Vollenhovermeer.

Construction on the Noordoost Polder started in 1937. In 1940 the last gap in the polder dike was closed. The main channels of the polder were dug while the polder was still submerged. It was pumped dry in 1942 (during the time of German occupation) after pumping 1.5 billion cubic meters (53 billion cubic feet) of water. This created 480 square kilometers (185 square miles) of new land. The deepest canals in this polder lie 5.5 meters (18 feet) below the IJsselmeer (van Duin and de Kaste 1990).

Soon after pumping this region dry, the engineers discovered that the old land began to dry more than was previously expected (based on experience with the Wieringermeer Polder). Prior to the completion of the Noordoost Polder, the groundwater levels in the vicinity of seashore towns like Kuinre (see **Figure 5-6**) were controlled by the water level in the Zuiderzee. After the Zuiderzee water was pumped out of the Noordoost Polder, the water table along the old shoreline was lowered to a level well below that of the former sea bottom. Calculations revealed that a marginal lake at least 2 kilometers (1.2 miles) wide would have prevented this from happening.

The Noordoost Polder had more heterogeneous soils than Wieringermeer. As a result, more work was required to prepare the soil for agricultural uses. Since the development of the Wieringermeer Polder, better equipment was developed for mixing the upper soil layers to provide a better soil profile. Large areas were deep plowed to a depth of 2 meters (6.6 feet), bringing up lower clay layers and covering the surface layer of sand. The plots in the Noordoost Polder were each 24 hectares (60 acres), making them somewhat larger that those of the Wieringermeer Polder. The land was leased to the farmers. The type of soil determined how the land would be used. Clay soils in the middle of the polder were used for arable farming. Lighter soils were used for grazing. Sandy soils were planted with trees for stands of woodlands.

Because of a shortage of construction workers at the end of World War II, new methods were needed to construct the farm buildings. Prefabricated concrete panels were used, giving this part of the IJsselmeer region a drab, uniform look.

Two islands were incorporated into the Noordoost Polder, Urk and Schokland. Urk was a very important center of the fishing industry prior to the construction of the Barrier Dam. It was feared that turning the Zuiderzee into a freshwater lake would destroy its economy. This has not happened. Urk fishers were able to sail through the locks and continue to fish in the North Sea. Today much of the fish is brought on land elsewhere but still shipped by truck to Urk for eventual sale. Urk is still one of the most important fishing towns in Northern Europe. Today, this former island is part of the mainland. When driving through the Noordoost Polder it is visible miles away due to its higher elevation. The contrast between the town itself, occupied since the seventeenth century, and the drab, prefabricated, post–World War II construction in the rest of the polder is very striking.

Schokland was not as fortunate as Urk. The island of Urk was a large deposit of boulder clay, while Schokland was an island in a peat landscape. Schokland was densely populated at one time. As the sea levels rose, it continued to erode until the inhabitants were evacuated in 1859 by royal decree. The island was abandoned. Today, this narrow rise in the flat polder landscape is a park and museum. There you can see the remains of a seawall that is no longer in sight of the sea.

Town planning in the Wieringermeer Polder was somewhat experimental and mostly unsuccessful. More effort was made in advance to plan the towns in the Noordoost Polder. Emmeloord is the main city, located in the center (see **Figure 5-6**). Emmeloord was to provide the administration as well as the business that would service the entire polder. There is a ring of ten villages surrounding Emmeloord (not shown in **Figure 5-6**). These were built for farm services and residences for farm employees. These villages were located in such a way that no one in the polder was more than 5 kilometers (3.1 miles) away from a town. This was deemed necessary for good community life, especially in an era in which the primary means of transportation was by bicycle.

By the time the Noordoost Polder was completed and the towns constructed, things had changed considerably. The post–World War II economy left fewer people on the farms and therefore fewer people to occupy the ten villages. The agricultural needs became more complex, meaning that the small villages could not support the agricultural community. Additionally, mopeds and automobiles became more common, eliminating the need for the tight village spacing. The smaller villages did not develop as planned. These lessons learned were applied to town planning in the next two polders.

The Flevoland Polders

The largest of the IJsselmeer polders is really two connected polders. Oostelijk (Eastern) Flevoland and Zuidelijk (Southern) Flevoland were constructed at different times but share a single dike boundary, called the Knardijk, effectively making them a single polder. The marginal lakes that surround these two polders (Gooimeer, Eemmeer, Nuldernauw, Wolderwijd, Veluwemeer, Drontermeer, and Ketelmeer) were included in the design to keep the former coastal land from drying out as the water level in the polder dropped. The result was the creation of a polder that was, effectively, a complete, two-cell, bathtub-like island in the IJsselmeer.

Work began on these polders much later than originally planned. This was primarily due to the reconstruction

Figure 5-16: Land use allocation in the IJsselmeer polders.
Data from IDG 1994.

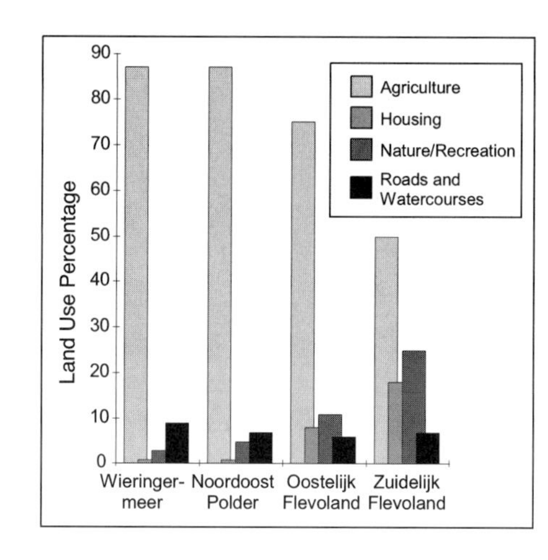

efforts that took place after World War II. Work began on Oostelijk Flevoland in 1950. The final closure of the dike occurred in 1956. By June 29, 1957, after the removal of 1.6 billion cubic meters (57 billion cubic feet) of water, the polder was dry, providing 540 square kilometers (208 square miles) of new land. Zuidelijk Flevoland's construction began in 1959 with final closure in 1967. The polder was dry by May 23, 1968 after pumping 1.4 billion cubic meters (40 billion cubic feet) of water. This polder added another 430 square kilometers (166 square miles) of new land (van Duin and de Kaste 1990). With the completion of the Flevoland polders, the amount of new land created between 1930 and 1968 was 1,650 square kilometers (640 square miles). This is roughly equal to 60 percent of the land area of the state of Rhode Island. For the Netherlands, this amounted to a 5 percent increase in dry land. It is also important to note that this increase in land was located near the main urban population centers.

The original plan for Oostelijk Flevoland, like the Noordoost Polder, included many more villages than were actually constructed. With increased mechanization, the farm sizes increased and subsequently the rural population density decreased. The number of cars increased, allowing greater distances between towns. Eventually only four towns were constructed in Oostelijk Flevoland. Dronten was constructed to be the central town in this polder. Originally ten small support villages were planned. Only two were constructed, Swifterbant and Biddinghuizen. The largest city built in Oostelijk Flevoland was Lelystad (see **Figure 5-6**), named after Cornelis Lely. Lelystad was designed to provide support to the entire IJsselmeer reclamation region. The projected population of this town was planned to be 100,000 by the end of the twentieth century. In 1986, when the Dutch government established the new Flevoland Province (including the Noordoost Polder and both Flevoland polders), Lelystad was designated the capital city.

As previously discussed, increased mechanization resulted in larger farm plots. The standard plot size was scaled up again to 30 hectares (74 acres). The standard plot was 1,000 meters (3,280 feet) long by 300 meters (980 feet) wide. Along the eastern edge the soils were sandier. Since the land is well below the level of the surrounding lakes, a sandier soil results in more upward seepage, thereby requiring a denser drainage network. As a result, the parcels along the eastern edge of the polder were smaller. The actual farm sizes were multiples of the 30-hectare (74-acre) standard plots.

Flevoland—Adapting to the Needs of Society

The first plan for the reclamation of the Zuiderzee surfaced in the seventeenth century. Plans that were constructible were first published in the nineteenth century. Construction began early in the twentieth century. The work needed to create dry land from sea bottom took nearly 50 years (1920 to 1968). Development of the newly reclaimed Flevoland polders continues into the twenty-first century. Clearly, over such an extended period of time, the needs of the country have changed, especially relating to uses for new land. The Zuiderzee reclamation plan has been continuously updated and revised to reflect the changing needs of society.

The last Zuiderzee Polder, Zuidelijk Flevoland, experienced the greatest changes from Lely's original concept plans to the current realization. As technological advances in farming increased, the need for new agricultural land decreased. Zuidelijk Flevoland is the polder closest to the major urban centers. The bridge entering the polder from the south is only 18 kilometers (11 miles) from the center of Amsterdam. The focus of the Zuidelijk Flevoland development shifted from agriculture to housing and recreation/nature. The land use allocations for all four Zuiderzee/IJsselmeer polders are shown in **Figure 5-16**. By the time Zuidelijk Flevoland was realized, agricultural allocations shifted from 87 percent in Wieringermeer and Noordoost Polder to 50 percent in Zuidelijk Flevoland.

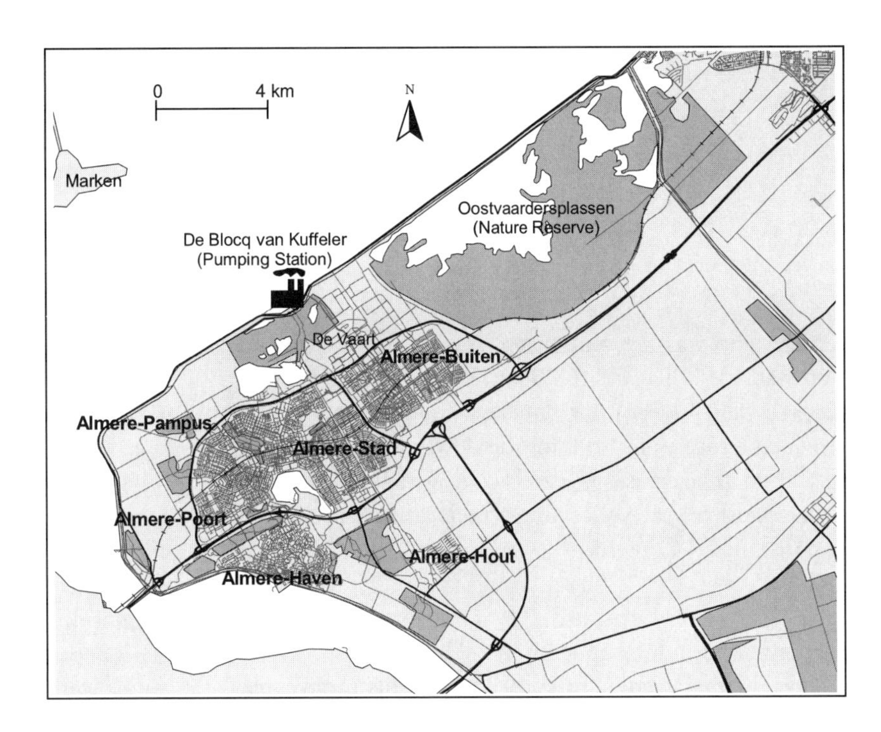

Figure 5-17: Almere and the Oostvaardersplassen.

For each of the Zuiderzee polders it was critical to have a good land use plan in place once the polder became dry. These plans laid out the location and number of towns that would be built and the organization of the agricultural land, including plot size and drainage. Decisions regarding the optimal use of the agricultural lands were based primarily on available soil types. The land use plans set up the infrastructure that gave the newly created land the support it needed to thrive.

The layouts of the larger cities have provided some of the greatest challenges in planning. Most city planners and municipal engineers work within an existing city, making plans and designing changes to existing city layout and infrastructure. Rarely have city planners and engineers had the opportunity to design and construct entire cities "from scratch." The largest and newest of the new cities planned and still under construction (in 2005) is Almere (see **Figure 5-6**).

Almere

Almere is the IJsselmeer polder town closest to the major urban population center in Amsterdam. For all practical purposes it is the last of the new cities in the IJsselmeer polders. It is probably the ultimate city planner's dream. In 1968 the planners and engineers were given the goal of designing and building a new city, starting with only dry ground.

The population projections for some of the earlier polders were often inaccurate. This was partly due to the length of time between the first plans and the final implementation. Since Almere was going to be the largest city in the IJsselmeer polder region, it would take several more decades for the development to be complete. A design concept that is developed for an entire city at one time may not be acceptable to new inhabitants arriving decades later. Eventually the city planners decided to construct Almere as a polynuclear city (see map in **Figure 5-17**). The larger Almere city was really several smaller cities or nuclei held together by a green space. Each nucleus had its own neighborhoods, central business district, and unique character. Each nucleus had well-planned public transportation systems including bus-only roads linking neighborhoods with the central business district and train stations. Each area had a well-planned system of bicycle and pedestrian paths that minimized contact with automobiles.

It was essential to planners that Almere not become simply a commuter suburb to Amsterdam, so they planned industrial zones and an attractive downtown. The city now consists of the three nearly complete nuclei—Almere-Haven, Almere-Stad, and Almere-Buiten. Two new nuclei are under construction or being planned, Almere-Poort and Almere-Hout.

In 1976 the first inhabitants moved into the 460-hectare (1,140-acre) Almere-Haven. Almere-Haven was designed for 20,000 inhabitants. This nucleus is distinctive in that it is situated along the Gooimeer (Lake Gooi) and has a central business district with access to the harbor.

Almere Stad was the central nuclei. The site for Almere-Stad was 2,500 hectares (6,200 acres) and included a sand pit lake called Weerwater (literally "water again"), which provided a nice waterfront for the downtown. In addition to its residential areas, Almere-Stad's downtown will feature high-rise office space, shopping centers, and many cultural opportunities. Construction began on Almere-Stad in 1979.

The 1,500-hectare (3,700-acre) Almere-Buiten is a residential area located between Almere-Stad and the Oostvaardersplassen nature reserve. Its distinctive features include an open landscape, a well-designed network of green space zones, close proximity to the De Vaart industrial zone, and unique architecture. Construction began in Almere-Buiten in 1983.

The population of the greater Almere city reached 100,000 by the mid 1990s. The city planner's goal is to reach a population of 180,000 by the time Almere-Stad is completed in 2006. The growth of Almere is significant compared to the other towns in the two Flevoland polders. The population of Lelystad stagnated around 60,000—its population at the start of Almere's growth. Almere is expected to continue to grow with the development of the two new nuclei, Almere-Hout and Almere-Poort, reaching 350,000 to 400,000 inhabitants by the year 2025.

The new downtown center of Almere-Stad is expected to be completed by 2006. The design for this part of the city allows easy access by automobile or public transportation but provides a pleasant environment for those living and working in this part of the city. All traffic, parking facilities, and many shops will be at the ground level. The rest of the shops and building entrances will be located on a curved, pedestrian-only deck that creates a second "ground" level. This deck starts at the lakeshore and gradually rises to a height of 6 meters (20 feet) above the ground surface. The slope of the deck will be gradual enough to make it hardly noticeable for those walking on this upper level. The lower level will be lit with daylight shafts, and its ceiling will be tall enough to allow busses to enter and to prevent it from feeling like a parking garage. The upper level will be a pedestrian zone free of traffic. The planners and engineers are making sure that this area is designed to be an attractive place to live and work.

Oostvaardersplassen—
A Nature Preserve below Sea Level

The deepest part of the former Zuiderzee in the area of the Flevoland polders was the Oostvaardersdiep. The Oostvaardersdiep runs along the northwest dike between Almere and Lelystad. When Zuidelijk Flevoland was declared dry in 1968, it really was not completely dry. The Oostvaardersdiep still had a large area of open water (see map in **Figure 5-17**). A couple of drainage ditches would have easily taken care of this. At that time, the plan for the polder called for this area to be used for industrial development. The necessary drainage works were simply postponed until the start of that development.

One of the first steps in curing the newly reclaimed land was sowing reeds from the air. The area along the Oostvaardersdijk was too wet for the reeds to take hold. Instead mud flats and lakes developed, covered with natural pioneer vegetation. Because the area was not developed right away, plants and animals had a chance to establish themselves. By 1973 the planners realized that the lake and marshland were far more valuable as a nature preserve than as an industrial site. Soon measures were being taken to ensure that this site would develop as a natural environment (Wigbels n.d.). These include the following:

- In 1974 an embankment was built around the area to ensure that water would not drain from the site.
- In 1976 pumps were installed to provide a water supply to keep the site from drying out under natural evaporation conditions. The water level in the marshland was now under total control. The levels could be optimized for the specific needs of the wildlife. Unfortunately, the conditions were too optimal. The goose population increased to the point where overgrazing and subsequent wind-driven erosion became a problem.
- Adding dry land to the nature preserve solved the goose problem. In 1982 a railroad line that was planned to run between Lelystad and Almere was built around the preserve, and the marginal land between the railroad and the marshes was added to the preserve (see map, **Figure 5-17**). This combination

of wet and dry lands allowed a complete marsh ecosystem to develop.

- A new effort to allow the area to develop naturally began in 1996. Water levels within the preserve will be altered only if they exceed certain extreme limits.

The Oostvaardersplassen has become a wildlife sanctuary of international importance.

The Markerwaard

In 1891 Lely proposed a large polder directly north of Amsterdam. In Lely's plan, shown in **Figure 5-3**, it was called the Z.W. Polder (or South West Polder). In later plans this polder was renamed the Markerwaard. **Table 5-1** tells the story of the Markerwaard. Construction started several times but was never completed. **Figure 5-6** shows dikes linking the island of Marken with the mainland as well as a dike connecting Enkhuizen with Lelystad. With the experience gained in reclaiming former sea-bottom land, the Markerwaard would be an easy project to complete.

The completion of the Markerwaard was often discussed. Those in favor saw benefit in new agricultural lands, new areas for urban expansion, new land for recreational use, and a location for a second national airport. Those against the project emphasized the valuable natural resource in the existing freshwater lake. As the discussions continued, the size of the proposed Markerwaard shrank. In 1991 the government finally decided that the Markerwaard would simply not be reclaimed. It remains today as the Markermeer, a freshwater lake.

Water Control

Figure 5-6 shows the water control points within the IJsselmeer region. Water levels can be separately controlled within the IJsselmeer, Markermeer, and two of the Flevoland boundary lakes, Veluwemeer and Nuldernauw Wolderwijd. The control is a function of the volume of water entering via precipitation, seepage, and river inflow. The control plan also accounts for evaporation losses. The largest volume of water enters the IJsselmeer through the IJssel River near Kampen at the gap between the Noordoost Polder and Oostelijk Flevoland. The IJssel River is a branch of the Rhine. Weirs constructed on the Lek (another Rhine River branch) control the volume flowing to the IJsselmeer (see Chapter 6).

The managed water levels in the IJsselmeer are different in summer and winter. In the winter the main goal is to adequately drain the surrounding old land. The prescribed water level was set at the low tide level of the former Zuiderzee—0.4 meter (1.3 feet) below NAP. This is the level Lely originally proposed in 1886. The lake level is allowed to rise 0.2 meter (0.7 foot) in the summer to allow easier irrigation in the surrounding old land (van Duin and de Kaste 1990). The IJsselmeer also acts as a large storage basin. Enough storage capacity is left to allow water to accumulate during times of high water levels on the Waddenzee.

Managing New Threats—Inflatable Dam at Ramspol

Closing off the former Zuiderzee from the North Sea significantly reduced the potential for flooding in the new IJsselmeer region, but a flaw in the design layout of the Noordoost Polder and Oostelijk Flevoland produced a new threat. When a storm produces strong winds from the west, the waters of the IJsselmeer are pushed into the gap between these two polders. The lake in this gap is called the Ketelmeer. The storm pushed water meets flows heading into the IJsselmeer from the rivers IJssel and Meppelerdiep (draining large parts of the provinces Overijssel and Drenthe). Water is pushed further past the narrows between the Noordoost Polder and the mainland at Ramspol and into the lake called the Zwartemeer (see map in **Figure 5-6**). This flaw has caused severe flooding along the Zwartemeer coast in the mainland.

After the high water levels in 1993 and 1995, the central government established the "Delta Plan Large Rivers." This plan placed a high priority on solving the flooding problems at Ramspol. The protection of the area around the Zwartemeer was the responsibility of the regional water board, Waterschap Groot Salland. Rijkswaterstaat (the central government water authority) was hired as the consulting engineer. (See Chapter 9 for more

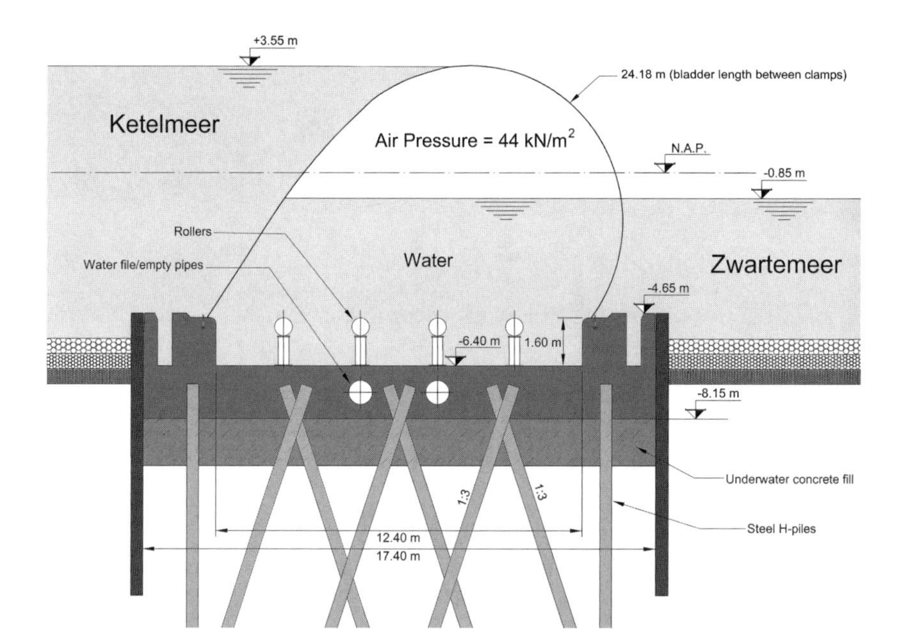

Figure 5-18: Cross section of the inflatable dam at Ramspol.
Drawing provided by Rijkswaterstaat Bouwdienst.

information on the regional water boards, Rijkswaterstaat, and "Delta Plan Large Rivers.")

Waterschap Groot Salland and Rijkswaterstaat considered several alternatives for the protection of the Zwartemeer. One option was to raise the level of all of the dikes in that part of the country. This option would be quite expensive and would require significant modification of the surrounding landscape and cultural/historical features. Another option used successfully elsewhere would be to build a storm surge barrier at the narrows between the Ketelmeer and the Zwartemeer at Ramspol. A storm surge barrier is a dam that is normally open to allow the flow of water and to permit shipping. When a storm threatens, the barrier is closed, thereby providing protection. This plan would still require dike raising on the Ketelmeer side of the barrier.

Several storm surge barriers had already been constructed in the Netherlands at the time this project was started in 1997. (See *The Delta Plan Overview* section in Chapter 8.) The barriers constructed have all been very expensive and involved building large structures that have a negative visual impact on the landscape. After a review of 34 different types of structures, they decided to construct an inflatable dam. In essence, this dam would be a very large rubber balloon normally hidden on the channel bottom. When a storm threatens, it is filled with both air and water, providing a protective barrier from flooding.

Inflatable dams are not a new concept. At the time of the design of the Ramspol barrier there were 2,000 in existence, 90 percent of which were found in Japan. The application of the inflatable dam technology in the Netherlands presented a significant design challenge for several reasons. First, it is the largest inflatable dam in the world. It consists of three sections, each 75 meters (246 feet) long, 13 meters (43 feet) wide, and 8.35 meters (27 feet) high (van der Horst n.d.). The Adam T. Bower Dam on the Susquehanna River at Sunbury, Pennsylvania, is the world's longest inflatable dam, using six 91 meters (300 feet) long rubber cells, but when inflated it creates a lake only 2.4 meters (8 feet) deep. Second, the Ramspol Dam is unique in that it is filled with both water and air. Some inflatable dams use water, some use air, but none use both. Finally, all other dams are intended for controlling water levels, as is the case of the Adam T. Bower Dam, which is designed to provide adequate depth for recreational boating. The Ramspol Dam is designed to be a storm surge barrier to be activated only during times of large storms. As a result, it needed to be designed to inflate quickly and withstand the forces associated with a large storm.

Figure 5-18 shows a cross section of the barrier. Under normal conditions, the balloon (really a rubber sheet attached to a concrete sill) lies hidden below the water surface, allowing ships to pass by unhindered. When flood protection is needed, air is pumped into the balloon to begin the inflation process. Also, upstream valves are opened, allowing water to flow in without requiring large pumps. After the storm passes, the water and air are pumped out of the balloon. Using a combination of both water and air resulted in a design requiring smaller pumps. The Ramspol Dam is designed to

▾ **Figure 5-19:** Rubber sheet being installed in the inflatable dam at Ramspol.
From Waterboard Groot-Salland.

◂ **Figure 5-20:** Two of the three sections of the inflatable dam at Ramspol.
From Waterboard Groot-Salland.

inflate in 60 minutes and deflate in three hours. Once it deflates, it is guided into a folded position by rollers located in the sill. The sill is supported by pile foundations driven to a depth of 15 to 30 meters (49 to 98 feet).

A number of design challenges delayed the construction of this project several years. The balloon is subject to extreme stresses, especially at the connections at each end, and thus had to be reinforced. Choosing the type of reinforcing fiber to use was a difficult decision to make. The balloon is made from 11 layers of rubber 16 millimeters (0.63 inch) thick, and is reinforced with nylon fibers. The original plan called for aramid fibers, but the higher elasticity of the nylon won out over the higher strength of the aramid.

The connections at the abutments were especially challenging. The Ramspol barrier is not really a balloon but a rubber sheet attached to the sill around its perimeter. The attachments along the length of the sill were not a problem. The problem occurred at the sloped abutments, where the length along the abutment wall was shorter than the expanded balloon length. Clamps were designed to take up the extra length in waves or folds. Detailed Finite Element Method calculations, scale modeling, and full-scale tests of the clamps were performed. The design constraints were set at a probability of failure of any part of the system of 0.00035 per year. The structure was designed to withstand the 10,000-year storm (van der Horst n.d.). The rubber sheet is designed for a life of 30 years, assuming that it is inflated twice per year.

The project was undertaken using a "Design-Build" contract. The contract included a ten-year maintenance agreement. Through this contract, the contractor benefited if the maintenance requirements were minimized by design. **Figure 5-19** shows the rubber sheet being installed. **Figure 5-20** shows two of the three sections of this dam inflated during a test performed on September 24, 2002, during which it operated successfully. An official opening of the structure was planned for December of the same year. Even before these official ceremonies could be held, the barrier went into action. This happened when a strong storm occurred on October 27, 2002, pushing water to 0.50 meter (1.64 feet) above NAP. This should be compared to the normal water level, which is between 0.4 and 0.2 meter (1.32 and 0.66 feet) below NAP. The barrier inflated automatically in response to the threat and worked successfully.

Places to Visit

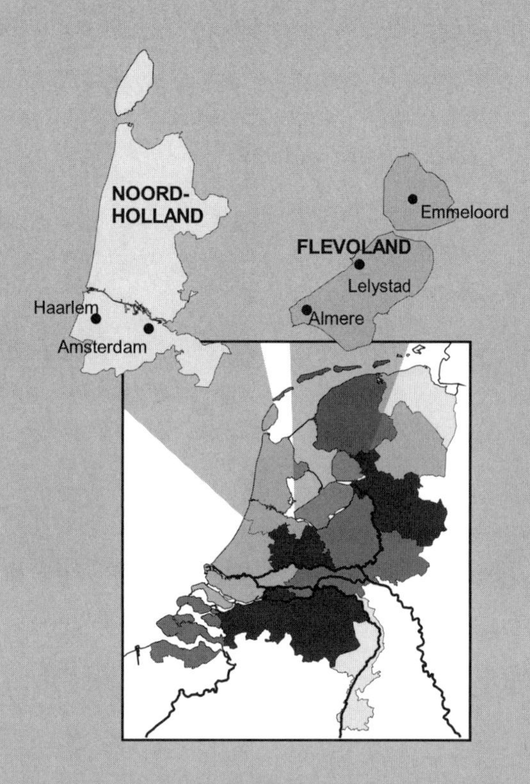

Zuiderzee Museum

DESCRIPTION: This exceptional museum provides a glimpse of the cultural-historical legacy of the Zuiderzee area prior to the construction of the Barrier Dam and the development of the IJsselmeer polders. It includes both an indoor segment and a large outdoor segment. It is located in the attractive old seaport town Enkhuizen.

LOCATION: In the city of Enkhuizen about 18 kilometers (11 miles) east of the city of Hoorn in the province of Noord-Holland.

INTERNET: www.zuiderzeemuseum.nl (Full information in English).

EMAIL ADDRESS: netpost@zuiderzeemuseum.nl.

ALONG EXCURSION: None.

New Land Heritage Center

DESCRIPTION: The New Land Heritage Center (Nieuw Land Erfgoedcentrum) has an extensive collection on the history of the Zuiderzee region as well as on the construction of the IJsselmeer polders. An informational film is available in English.

PHONE: 0-320-260-799 (from the United States 011-31-320-260-799).

ADDRESS: Oostvaardersdijk 01-13, 8242 PA Lelystad.

LOCATION: On the west side of Lelystad in the province of Flevoland. The center is near the point where N302 starts the path across the IJsselmeer to Enkhuizen.

HOURS: Tuesday through Friday, 10:00 a.m. to 5:00 p.m., Saturday and Sunday, 11:30 a.m. to 5:00 p.m.

ENTRY FEE: € 7.00 (free with Dutch Museum Card).

INTERNET: www.nieuwlanderfgoedcentrum.nl (limited information in English).

EMAIL ADDRESS: info@nieuwlanderfgoedcentrum.nl.

ALONG EXCURSION: Zuiderzee Reclamation.

Oostvaardersplassen Nature Reserve

DESCRIPTION: This 5,600-hectare (14,000-acre) nature reserve is in the lowest area of the Zuidelijk Flevoland Polder. It developed as a nature reserve almost by accident. There are extensive trails for hiking and bicycling and three wildlife observation posts. The information center features some English language materials.

PHONE: 0-320-254585 (from the United States 011-31-320-254585).

ADDRESS: Kitsweg 1, 8212 AA Lelystad.

LOCATION: Between Almere and Lelystad in the province of Flevoland.

HOURS: Information center: April through October: Tuesday through Sunday, 10:00 a.m. to 5:00 p.m.; November through March: Saturday and Sunday, noon to 4:00 p.m.

ENTRY FEE: None.

INTERNET: None.

ALONG EXCURSION: Zuiderzee Reclamation.

Schokland

DESCRIPTION: This is the former island now surrounded by land of the Noordoost Polder. The site includes a small museum, the church that was one of the few structures remaining after the island was abandoned, and a short section of sea wall.

PHONE: 0527-251396 (from the United States 011-31-527-251396).

LOCATION: Along N352 about 12 kilometers (7.5 miles) east of Urk in the province of Flevoland.

HOURS: April through October: Tuesday through Sunday, 11:00 a.m. to 5:00 p.m.; November through March: Friday through Sunday, 11:00 a.m. to 5:00 p.m.

ENTRY FEE: € 3.50

INTERNET: www.schokland.nl (limited information in English).

EMAIL ADDRESS: info@schokland.nl.

ALONG EXCURSION: Zuiderzee Reclamation.

Storm Barrier at Ramspol

DESCRIPTION: The barrier dam at Ramspol is the world's largest inflatable dam. It is one of the few used as a storm barrier. The south end of the structure is used as a visitor's center by special arrangement only. According to the Web site, weekday visits can be arranged for groups of ten or more by calling 038-455-7219 (from the United States 011-31-38-455-7219).

LOCATION: About 8 kilometers (5 miles) north of the city of Kampen along N50, at the border between the provinces of Flevoland and Overijssel.

ENTRY FEE: None.

INTERNET: www.wgs.nl is the Web site for Waterboard Groot Salland (primarily in Dutch). Once there, look for an English language switch. If a link to the barrier page (in English) is not found, then do a search ("zoeken") for "Ramspol Barrier."

ALONG EXCURSION: Zuiderzee Reclamation.

Barrier Dam

DESCRIPTION: The road along the Barrier Dam has a small rest area with a viewing tower. The rest area has a small restaurant and gift shop. The tower provides a good view of the dam. The restaurant décor includes many photographs of the construction of the dam.

LOCATION: Highway A7 between the provinces of Noord-Holland and Friesland.

ALONG EXCURSION: Frisian Coast and Dwelling Mounds.

6 The Rivers

Two large European rivers, the Maas and the Rhine, converge in the central portion of the Netherlands. Modifications to this great river region were made as early as the Roman period. Many major changes were made to discharge routes in the second half of the nineteenth century. The twentieth century saw further improvements in water management, flood control, navigation, and land consolidation.

The Rhine River and Maas River are important to the hydrology of the Netherlands. The surface water entering the country today is 25 percent from precipitation and 75 percent from rivers flowing into the Netherlands from other countries. All but a small fraction of the river water flow into the Netherlands comes from the Rhine and the Maas. **Figure 6-1** shows the contributing watersheds for these two rivers. The Maas (and the Schelde) rivers provide drainage for most of Belgium and a small part of France. At the town of Lith near the confluence with the Rhine, the Maas's drainage area is 29,000 square kilometers (11,200 square miles). The Rhine River provides drainage for a large part of Germany, nearly all of Switzerland, and a small part of Austria. More than half of the Netherlands lies in the Rhine watershed. The total Rhine catchment is 185,000 square kilometers (71,400 square miles). Its total length is 1,320 kilometers (820 miles).

The seasonal variations in flow for these two rivers are quite different, as can be seen in the inset graph in **Figure 6-1**. The Maas is a rain-fed river. The ratio of monthly average high to monthly average low is about 4.5. The Rhine River is fed by both rainfall and alpine meltwater. This results in a more uniform seasonal variation, since the peak meltwater period is in June. The ratio of monthly average high to low is only 1.7.

These two rivers are also quite different in terms of their branches. The Maas has only a few branches in the Netherlands, whereas the Rhine splits into many branches in the Netherlands. Most of the rivers mentioned in this book are branches of the Rhine.

Early Modifications along the Rhine and Maas

Early Configuration of the Rivers

Figure 6-2 shows the configuration of the Maas and Rhine between 800 CE and 1250 CE. After entering the Netherlands, the Rhine splits into two branches. The northern branch kept the name Rhine (sometime referred to as the Lower Rhine or Neder Rijn in Dutch) while the southern branch was called the Waal. Several significant rivers branched off the Rhine after the split with the Waal. The Vecht and the IJssel flowed north into Lake Almere (in 800) or the Zuiderzee (in 1250) The Lek flowed to the southwest toward the current city Rotterdam. The Old Rhine (Oude Rijn in Dutch) flowed to the North Sea at Rijnmond near the current city of Leiden. The Maas and the Waal combined at Gorinchem and became the Merwede. The Merwede and Lek combined

▾ **Figure 6-1:** Contributing watersheds for the Rhine and the Maas. Values shown in boxes are the flood discharge in cubic meters per second on January 31, 1995.

Reprinted, by permission, from van de Ven 2004.

◢ **Figure 6-2:** Configuration of Rhine/Maas system—800 CE and 1250 CE.

Redrawn from SWAVN 1984, Fig. 6 and van de Ven 2004, Ch. 3, Fig. 1.

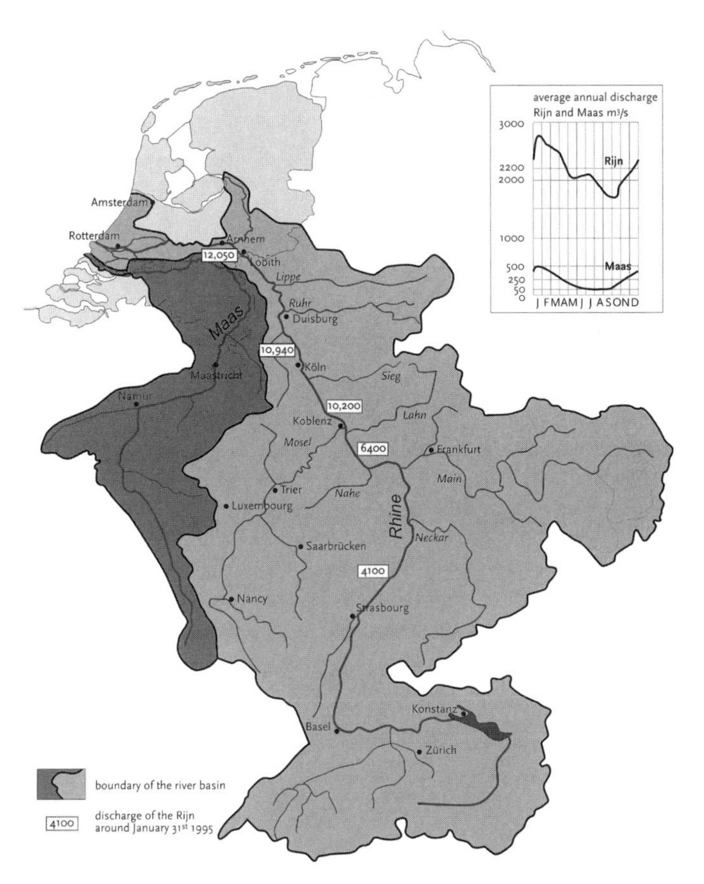

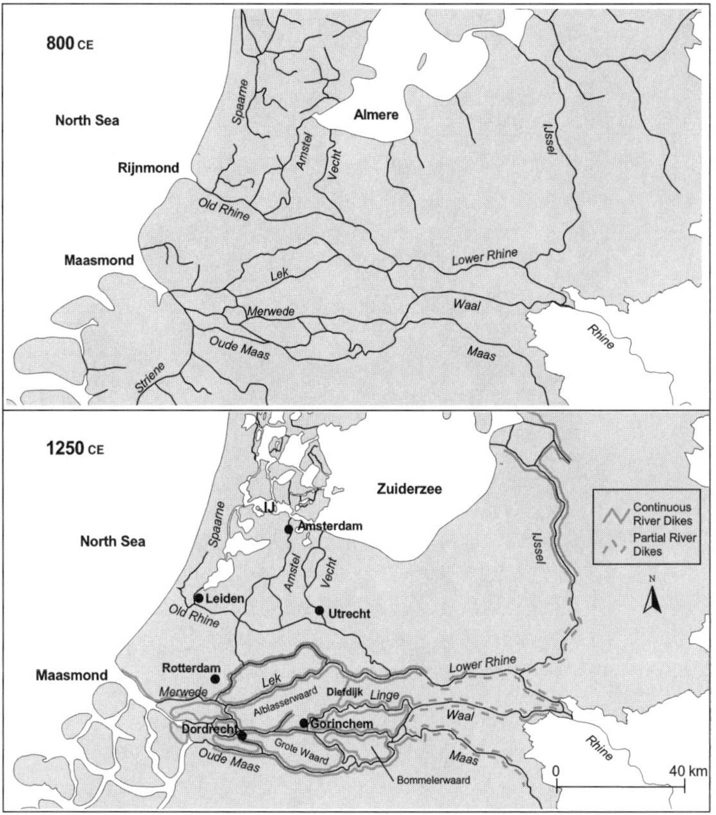

near the current city of Dordrecht and flowed to the North Sea at Maasmond. A branch of the Maas called the Oude Maas flowed to the sea at Maasmond.

Dike Construction

The first significant change made along the rivers was the building of dikes. This started around the year 1100, or 100 years after drainage modifications were started along the coast. The land that was settled along the rivers was barely above the normal high river levels. The farmers who settled here drained the land for their own use. To protect their investment, they also built low embankments on all sides. Water was discharged from the drainage ditches to the river through a sluice in the river dike. Gradually, more and more of the rivers had dikes built along their banks. By 1150 continuous dikes were built along the banks of the Lek and the Hollandse IJssel rivers. By 1250 the Oude Maas, and the Merwede were completely diked (see **Figure 6-2**). Damming of tributaries had already started by 1250 (van de Ven 2004).

All of this began a process that required the dikes to be continually raised over the following centuries. When one of the original low dikes failed, some of the land behind the dike would flood. If the flooded areas drained prior to spring planting, there would be minimal economic loss. But, with each flood event, the dikes were repaired and raised higher. As the dikes rose, the subsequent flooding events became more serious. With the land along the rivers protected from high water levels,

Figure 6-3: Formation of the Biesbosch.
Redrawn from Meijer 1996, Fig. 20.

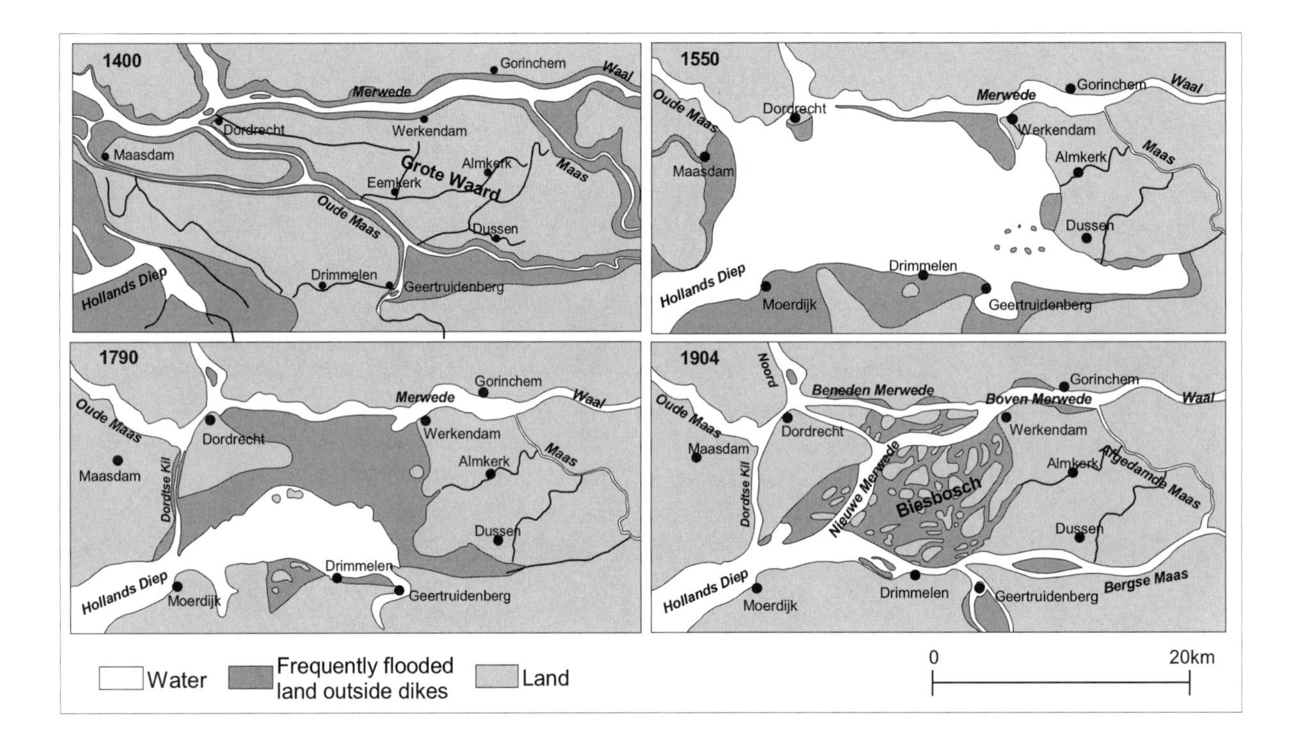

the silt and sand carried by the rivers could settle only in the rivers themselves. In addition, drainage of the land along the river caused it to subside. The sedimentation in the rivers also caused problems with winter ice. Ice became grounded in the shallow parts of the rivers. Additional ice was then pushed up behind the grounded ice, forming ice dams. These ice dams caused many dike breaches, especially in the period from 1600 to 1850. All of these factors created a river network prone to flooding and instability.

Changes at the Old Rhine Outlet

One of the major changes to the distribution of water through the Rhine delta was the silting up of the Old Rhine outlet at Rijnmond (now the location of the city of Leiden) between 1175 and 1200. This change occurred along with the formation of the young dunes as well as the storm surges that took place between 1163 and 1170. As a result, the area surrounding Leiden suffered significant drainage problems. Many measures and countermeasures were taken to alleviate the drainage problem. One measure was the placement of a dam on the Old Rhine at Zwammerdam (about halfway between the Vecht and the North Sea coast—see **Figure 6-2**) (van de Ven 2004). This drove more of the Rhine water north up the Vecht River. At the same time the drainage from the areas around Leiden was sent to the lakes to the north.

By the first half of the thirteenth century, the Zuiderzee and its arm, the IJ, had reached its greatest extent. In 1248 a storm on the IJ penetrated the lakes north of Leiden via the Spaarne. A dam with sluice gates was constructed there for the protection of these lakes. In later years these lakes continued to grow until they coalesced to form the Haarlemmermeer (see Chapter 4).

Around this time the Amstel River flowed south toward Utrecht. In the first half of the thirteenth century it connected with a northern flowing river that existed prior to the formation of the IJ. As a result, the Amstel turned northward. Between 1265 and 1275 a dam was built where the Amstel discharged into the IJ. This dam is now at the center of the city of Amsterdam.

Formation of the Biesbosch

At the beginning of the fourteenth century the individual dikes constructed along the rivers were connected to form a number of closed regions. Figure 6-2 shows some of these completely diked regions. These regions included Alblasserwaard between the Lek and the Merwede, Grote Waard between the Oude Maas and the Merwede, and the Bommelerwaard between the Maas and the Waal Rivers.

One of the more significant events in the lower reaches of the Maas and Waal network was the Saint Elisabeth's Day Flood of 1421 (already mentioned in Chapter 1). Figure 6-3 shows the Grote Waard around the year 1400. The Grote Waard was a region of around 200 square kilometers (77 square miles). It was a thriving agricultural area, containing 16 villages (Meijer 1996). One November night in 1421 a storm surge breached the dikes on the Hollands Diep, flooding the entire Grote Waard. The breached dikes had been weakened by peat extraction activities and lack of maintenance. One year later another flood breached the dikes along the Merwede near Gorinchem. The Grote Waard was never fully reclaimed and eventually turned into a unique natural environment. Because of water flowing in from the Merwede, the rivers of the former Grote Waard region remained fresh, but its proximity to the sea provided tidal fluctuations as high as 2 meters (6.6 feet). With time the sediment from the rivers began to fill in some of the flooded areas. The area then became a permanent freshwater tidal delta called the Biesbosch or reed forest. Figure 6-3 shows the change in the area from 1400 to the early twentieth century. Eventually some of the original Grote Waard was reclaimed, but the Biesbosch remains.

Two major changes occurred in this area in the twentieth century. Construction of dams along the North Sea as part of the Delta Project (see Chapter 8) reduced the tidal variations from 2 meters (6.6 feet) to 20 centimeters (8 inches) overnight (Meijer 1996). This was a major blow to the Biesbosch environment. Second, large storage reservoirs have been built in this area. Water is pumped out of the Maas River into these storage reservoirs. It is then pumped to water treatment facilities in the region for use as drinking water. There are two major benefits associated with the use of these reservoirs. First, the water is naturally purified by detention in the reservoirs. This reduces the amount of treatment required at the treatment plants. Second, water can be selectively removed from the Maas River. If the Maas water contaminants exceed certain limits, the pumps can be turned off for several weeks without impacting the drinking water supply to the region.

Local Drainage along the Rivers

The diked-in areas between the river branches presented some significant drainage challenges. The best route for interior drainage followed the downward slope of the land from east to west. This gave the interior drainage paths an opening to the rivers at the lowest possible elevations. The openings contained sluices for protection during high river levels. These diked-in areas could be compared to a bathtub slightly tilted from east to west. Excess rain in the winter along with high river levels would cause flooding in large areas for several months at a time. The flooding concentrated at the lower westerly end of these regions. To alleviate the flooding, some of these areas were provided with north-south running dikes or low embankments to keep all of the water from settling at the westerly end. One example is the Diefdijk shown in Figure 6-2.

As discussed earlier in this book, the seventeenth century was a time of great prosperity and a time when windmills were used extensively for drainage. This was true for the Noord-Holland drained lakes (discussed in Chapter 3), and it was true for this area along the Maas/Rhine Rivers. In Noord-Holland, storage basins were used to store the water from the drained lakes until it could be released to the sea. The storage basin concept was also put to use along the rivers.

Water management in the Alblasserwaard area (see Figure 6-2) was controlled by two water boards, the Overwaard and the Nederwaard. Drainage from both water boards was discharged to the Lek River at the western end of the Alblasserwaard near the village of Kinderdijk. The Nederwaard and the Overwaard each had its own canal, and they ran next to each other near the discharge point (see Figure 6-4). Drainage was not a problem when the water level on the Lek was low. Initially, during times of high water level on the Lek, the discharge canals could

▸ **Figure 6-4:** Discharge arrangement at Kinderdijk.

Reprinted, by permission, from van de Ven 2004

▾ **Figure 6-5:** Selected canals and the three weirs along the Lower Rhine River.

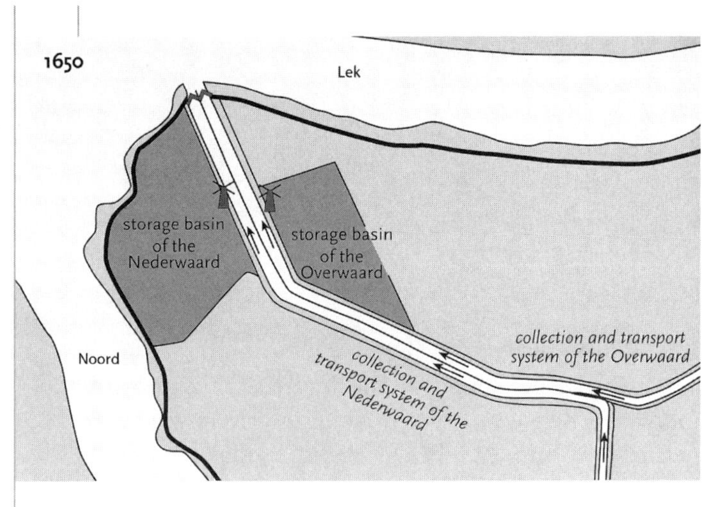

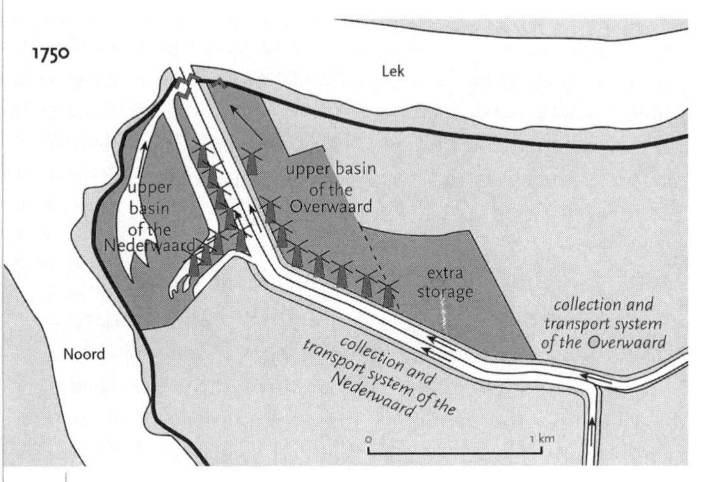

store the drainage water until the Lek levels dropped. As mentioned earlier, the diking-in of the rivers resulted in more frequent flooding. Water from the region needed to be stored for a longer period, and as a result a larger storage volume was needed. In 1366 the water boards purchased land on either side of the discharge canals for additional storage. When discharge into the Lek was not possible, water was diverted into these storage basins via a sluice. Once the levels on the Lek dropped, the basins were drained. Eventually, each basin was provided with one windmill, which served to drain the basins so that they could be used for pasture and hay growing during the summer (see the upper panel of **Figure 6-4**).

As the need for storage grew, a new approach was conceived. The dike encircling the storage basin was raised to provide greater storage volume. Next windmills were constructed to pump water from the discharge canal into the larger storage basins. These high basins could then discharge directly into the Lek even when the river water level was high. A flood event in 1726 inspired the water boards to begin raising the level of the storage reservoirs. In 1738 the Nederwaard constructed eight windmills, and in 1740 the Overwaard built eight windmills. The Nederwaard windmills were round and constructed of brick. The Overwaard windmills were octagonal with thatch covering a wood frame. The storage reservoir dikes were raised again in 1746 and 1766 (van de Ven 2004).

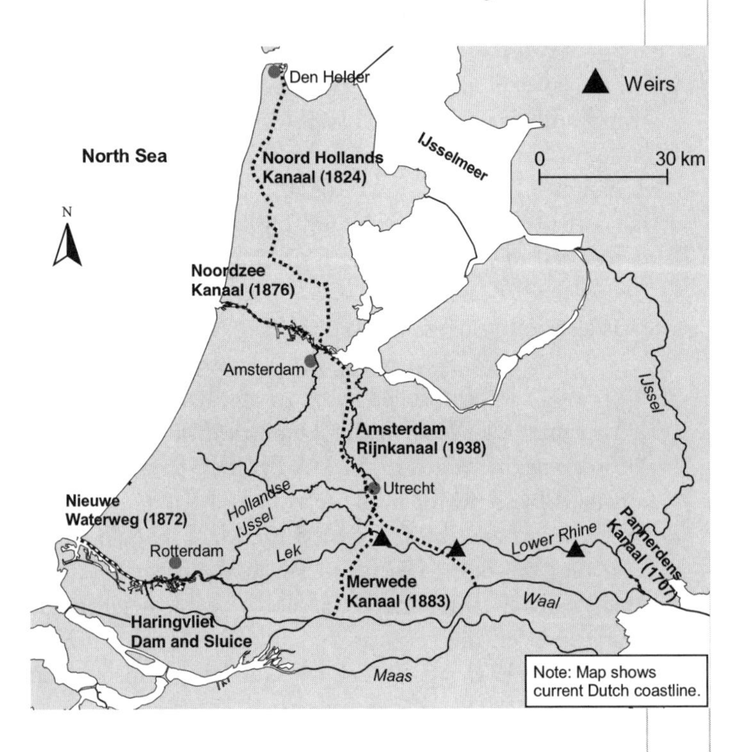

The collection of windmills draining the Alblasserwaard at Kinderdijk has been continuously maintained since construction. They were still in active use in the twentieth century, in particular during World War II. The site has been designated as a UNESCO World Heritage site. Nowhere else in the world can you find as high a concentration of working windmills.

Developments in the Seventeenth and Eighteenth Centuries

The area along the rivers presented a constant flooding challenge in the seventeenth and eighteenth centuries. As a defensive measure, people constructed dwelling mounds to be a place of

▸ **Figure 6-6:** Cross section of river showing summer and winter dikes.

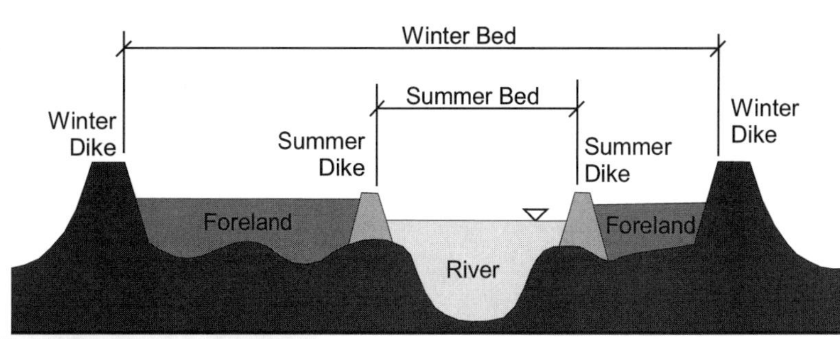

▾ **Figure 6-7:** River improvements along the Maas and the Waal.

Reprinted, by permission, from van de Ven 2004.

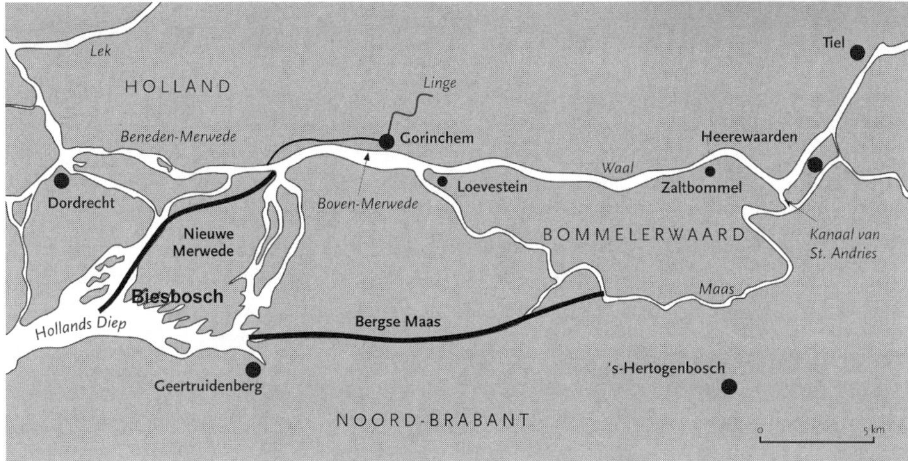

refuge during floods. Homes were also constructed with raised first floors. Some of the homes included an upper story to house animals in times of flooding.

By the end of the seventeenth century up to 90 percent of the Rhine River flow traveled down the Waal branch. This left the Lower Rhine and IJssel with only 10 percent of the flow. As a result, these branches began to fill in with sediment. To correct this problem, a new canal was constructed at the point where the Rhine and the Waal split. Along with other changes in the upper reaches of the Rhine River branches, this canal—called the Pannerdens Kanaal and shown in **Figure 6-5**—corrected the problem and returned the flows to a six-ninths Waal, two-ninths Lower Rhine, one-ninth IJssel split (van de Ven 2004).

River Modifications of the Nineteenth and Twentieth Centuries

By the middle of the nineteenth century, the situation along the rivers had become severe. Flooding and navigation problems had only gotten worse over the years. The following are some of the reasons for these problems:

- The rivers were poorly maintained. Many had sediment deposits that hindered flow and navigation. This created many islands and sandbars in the river beds.
- Dikes no longer followed a straight path along the river. When a dike burst, the water flowing through the hole formed a deep pool. When the dike was repaired, the new dike followed a route around the outside of the hole.
- The winter flood plain was not well designed or maintained. Along many reaches, there were two parallel dikes on each side of the river. There was a lower summer dike close to the river and a higher winter dike farther away. The area between the dikes, or foreland, was farmed during the summer and used as a flood plain in the winter. (See river cross section shown in **Figure 6-6**). In many areas the drainage path between the dikes was obstructed or the summer dikes were built too high to make the flood plain effective.
- There were too few discharge routes for the river water. The Waal and Maas joined at Gorinchem (see **Figure 6-7**). The confluence of these rivers was a location of significant flooding. Ice dams also regularly formed in this area.
- Upstream of Gorinchem (at Heerewaarden) the Maas and Waal were very close to each other. With no dikes at this location, the waters flowed over the small piece of land that separated the two rivers. Usually this increased the flow on the smaller Maas, causing flooding downstream.

Figure 6-8: Traffic on the Noord Hollands Kanaal.

- Too much water flowed through the Biesbosch, leaving the Merwede without enough flow. As a result, it silted up.

In the early part of the nineteenth century, studies were performed and plans were developed to correct the river problems. Among other results, these studies established the standard widths required for adequate floodwater conveyance. Because of the volatile political situation and lack of funds, no progress was made on river modifications until later in the century. The major improvements eventually made included the construction of two new branches, the Nieuwe Merwede and the Bergse Maas. These are shown in **Figure 6-7**. Construction of the Nieuwe Merwede began in 1850 and was completed in 1890. The Bergse Maas was completed in 1904. In addition, a dike was constructed at Heerewaarden to complete the separation of the Maas and the Waal. A lock was also constructed in the canal, linking the Maas and the Waal at this point (Kanaal van St. Andries). With these two river branches complete, work could then begin on improving the condition of the rest of the system. Giving the rivers the appropriate standard width was the primary activity.

Several improvements were made on the Lower Rhine and the Lek. Groins were constructed along the river banks to increase low flow depth, which would improve navigation. Two sharp bends in the river were also removed.

In addition to improvements to the rivers, several new canals were dug in the nineteenth century to provide better shipping access between the North Sea and the Dutch harbors. The sediment that no longer settled in the flood plains due to the construction of the dikes was filling in some of the Dutch harbors. By the eighteenth century a sandbar obstructed the entrance to the port at Amsterdam, limiting the water depth to nine feet. The Dutch ships had to be designed to be able to enter the silted harbors. Larger boats were either pulled or lifted through a shallow entrance. These boats could not be built for speed or size. This problem contributed to the decline in the Dutch trading industry during this period.

In 1815, after the end of the Napoleon era, King William I set out to solve the problem of silted-in harbors and sea entrances. First, he authorized construction of a new entrance to the harbor at Amsterdam. The engineers considered a direct cut to the sea through the dunes. This approach was rejected due to technical difficulties. Instead, a canal was dug from Amsterdam to the North Sea harbor at Den Helder (see map in **Figure 6-5**). This canal, called the Noord Hollands Kanaal, was constructed between 1819 and 1824—around the same time as the Erie Canal in the United States. The Noord Hollands Kanaal was designed and executed by Jan Blanken, one of the first leaders of the national water management agency, Rijkswaterstaat. It was an impressive project for its time. The canal was 79 kilometers (49 miles) long, 5 meters (16 feet) deep, and 37 meters (120 feet) wide. It included 14-meter (46-foot) wide locks at Amsterdam and Den Helder. It was constructed using shovels, wheel barrows, and mud mills—horse-driven dredges used as early as 1620 to dredge channels (van Veen 1955). Unfortunately, the canal was not a great success. By the time it was completed, it was already too small for the largest ships of the day. The many bridges and turns in the canal made for a slow trip. In the winter, travel was further hampered by ice. As a result, Amsterdam continued to decline as a center for trade. Today, the Noord Hollands Kanaal is in active use primarily for pleasure boats and boats transporting freight inland (see photo in **Figure 6-8**).

In 1862 Prime Minister Thorbecke set out to improve the entrance to the Rotterdam harbor. In addition to providing an open entrance to the harbor of Rotterdam, such a new waterway, if made large enough, could become a new outlet for the Rhine River branches. Engineer Pieter Caland suggested a unique construction

Figure 6-9: Rhine River visor weir in the open position.

method. His plan was to cut a small canal from Rotterdam to the North Sea. Then, after closing the wider mouth to the North Sea, the hope was that the water forced through this small cut would gradually widen the cut and create a deep waterway to the sea. His idea did not work very well. The initial cut through the dunes at Hoek van Holland was started around 1865. By 1872 the currents had cut enough to allow the first boat to sail through, but more work was needed to turn it into a successful shipping canal (van Veen 1955).

Once steam dredges capable of operating on the open sea were developed, work began on widening and deepening this new canal. After many years of continual dredging, this new canal, called the Nieuwe Waterweg (in English, New Waterway), reached the required depth in 1896. By 1909 its width was 100 meters (328 feet). This canal did not include locks at the North Sea and thus left an open connection between the North Sea and the port of Rotterdam. Because of the success of the Nieuwe Waterweg, the Rotterdam harbor expanded, making it the largest in the world.

Around this same time, Amsterdam improved its connection with the North Sea by constructing the Noordzee Kanaal, a canal that ran directly from Amsterdam to the west (see **Figure 6-5**). This canal was completed in 1876. In 1896 a new ship lock was completed that was 225 meters (738 feet) long, 25 meters (82 feet) wide, and reached a depth of 10 meters (33 feet) below NAP. To keep ahead of the growth in the size of cargo ships, a second lock was added in 1928 that was 400 meters (1,310 feet) long, 50 meters (164 feet) wide, and reached a depth of 15 meters (49 feet) below NAP. For many years this was the largest lock in the world (van de Ven 2004).

Further improvements were made to the Lower Rhine and the Lek rivers later in the twentieth century. After the construction of the Zuiderzee Barrier Dam in 1932 (see Chapter 5), it was clear that the country needed better control of freshwater distribution from the Rhine. In particular, during the summer months, higher flows were needed in the Nieuwe Waterweg to avoid salination of the rivers in the Rotterdam area. Doing this would threaten the freshwater supply to the IJsselmeer. This problem was solved by closing off other discharge points along the coast and by creating better flow control along the Rhine.

Coastal discharge was controlled with the completion of the Haringvliet Dam and sluice complex in 1971. Work along the Rhine and Lek included the canalization of several sections and the construction of three flow control weirs (see **Figure 6-5**). Work began in 1954. The upper weir was completed in 1970. This weir could be closed to force more water up the IJssel and on into the IJsselmeer. The lower two weirs were put in place to keep the Rhine and Lek navigable. They were located just upstream of the intersections with the Amsterdam Rijnkanaal and the Merwede Kanaal. All three weirs include locks for ship passage. During periods when water need not be diverted, the weirs are lifted out of the water, giving them the appearance of a visor on a cap. They are therefore referred to as a "Visor Weir." **Figure 6-9** shows one of these weirs in the open position.

Consolidation of the Land between the Rivers

In the region between the Maas and the Waal rivers, the land on the high natural river levees was suitable for arable farming. The lower-situated, poorly drained basin clay areas were good only for pasture and production of osier. Since the roads followed the levees, the lower areas also suffered from poor access. As a result, the region had poor drainage and few roads—many impassable when wet. For many centuries large parts of this area were flooded from November to May. All farmsteads were located in the dike villages for safety, and their plots were situated several kilometers away in the lowlands. Further

Figure 6-10: The Maas River during the 1995 flood. This photograph was taken in Wamel, about 35 kilometers (22 miles) west of the city Nijmegen.
From Pim Beukenkamp.

complicating the situation, because of long-standing inheritance rules, a given farmer's holdings were often divided into several narrow parcels widely distributed throughout the area. In the Bommelerwaard West district, for instance, the average holding was 7.7 hectares (19 acres) and split among more than six parcels, often as narrow as 20 meters (65 feet). The parcels frequently had no road access, requiring the farmer to obtain access easements (Meijer 1981).

To improve the situation, land consolidation projects were begun around 1950. The application for a complete redesign of the landscape in Bommelerwaard West was submitted in 1959, and the provisional plan was approved in 1964 and endorsed by the landowners in 1966. The plan included the following:

- The number of distinct parcels was reduced from 2,415 to 718, and the size of the average parcel increased from 1.2 hectares (3 acres) to 6.5 hectares (16 acres).
- Ninety-four kilometers (58 miles) of drainage canals were redug or widened.
- Fields were leveled and drainage tiles were installed.
- Thirty-two small sluices were constructed to improve water level control.
- Forty-four kilometers (27 miles) of third class roads were constructed or improved.
- Twenty-nine farms were relocated from the dike villages to the farmstead.
- Existing nature areas were protected and 12,000 trees and 395,000 shrubs were planted.

The work was completed by 1975. In November 1976 the reallocation of land was finally certified by a notary. Subsequently, similar projects were carried out in other regions in the river district.

Room for the Rivers—Planning for the Future

The improvements made to the rivers seemed to be quite effective. Between 1926 and 1993 there were no major river floods. The high water levels that occurred around Christmas 1993 did not initially raise major concern. Concerns about the condition of the rivers were raised when, two years later, the flooding returned. Late January of 1995 produced enough rain to cause flooding in Germany, France, Luxembourg, Belgium, and the Netherlands. By February 1, 1995 the flood water had reached the top of the winter dikes, causing mass evacuations in the region (see **Figure 6-10**). Furthermore, a long period of high discharge saturated the river dikes and put them in danger of collapse, causing heightened concern. The entire region between the Lower Rhine, Waal, and Maas rivers was evacuated. A quarter million people, 300,000 head of cattle, and millions of hogs and chickens were relocated for almost a week. The flooding problems of 1995 occurred upstream of the location where improvements were made at the turn of the twentieth century. Fortunately, none of the river dikes collapsed, and the danger diminished as the floodwaters receded.

The first task following these flooding events in 1993 and 1995 was to strengthen the dikes. The "Delta Plan Major Rivers" was established to strengthen existing dikes. Dike reinforcement and levee construction was done along 740 kilometers (460 miles) of river. This work, completed in 2000, was only a short-term correction. It became clear that a new approach was needed to solve the problems associated with river flooding.

Following the flooding in 1995, a policy document was put into effect. This document, called "Room for the River," seeks to find ways to expand the discharge capacity of the major rivers as well as to reduce flood damage.

One of the results of the 1993 and 1995 floods was that the design discharge for the Rhine River changed (RWS-DZH,DON,RIZA 2001). The design discharge used in the Netherlands is the discharge that is equaled or exceeded, on average, once every 1,250 years. The discharge database used to make the calculations is based on flows at Lobith (at the German border) measured since 1901. With only 100 years of data records, the 1993 and 1995 flood events had a significant impact on the design discharge calculations. As a result, the design discharge rose from 15,000 cubic meters per second (530,000 cubic feet per second) to 16,000 cubic meters per second (565,000 cubic feet per second). Hydraulic modeling calculations indicate that this will result in a 20 to 30 centimeter (8 to 12 inch) increase in the water surface in the upper reaches of the system. This design discharge is anticipated to increase further due to climate change. By 2100 it is expected to reach 18,000 cubic meters per second (636,000 cubic feet per second). The "Room for the River" document recognized that continuing to raise the river dikes, in response to increasing design discharges, was not a wise policy. This would simply increase the amount of damage that will occur if a breach happens. The main challenge recognized is to find ways to accommodate this increased design discharge without raising the dikes.

Accommodating this increase requires a combination of increased conveyance and storage. A wide range of options for accomplishing this goal have been considered and analyzed (RWS-DZH,DON,RIZA 2001). Plans are being made to implement the most effective set of river modifications.

Two areas have been identified as potential sites for large-scale emergency detention. The Rijnstrangen and Ooijpolder areas, located near the Dutch-German border, together are 45 square kilometers (17 square miles) and can provide as much as 215 million cubic meters (7.6 billion cubic feet) of storage. These areas are sufficient to lower the peak discharge by as much as 1,000 cubic meters per second (35,000 cubic feet per second). They would remain in use until the storage was needed. The Rhine basin is large enough that advance flood warnings are possible. Evacuation plans and damage compensation plans are needed to make these storage areas viable.

The excavation of the flood plain and removal of hydraulic bottlenecks (ferry ramps, bridge abutments, summer dikes, etc.) are other options that could have a significant effect on reducing flood levels. With flood plain excavation, the sediment that has accumulated in the foreland (see **Figure 6-6**) is removed. If the flood plains are excavated and if the summer dikes are left intact so that the land can be used for agricultural purposes, the flood elevations on the Waal could be reduced by as much as 50 centimeters (20 inches). If the excavation is more extensive and the summer dikes are removed, resulting in a wet area to be used for nature preserves and lakes, then a flood elevation reduction of 80 centimeters (31 inches) is possible on the Waal. Using water surface models, Rijkswaterstaat engineers predicted that the removal of the 60 worst hydraulic bottlenecks could reduce the flood elevation by 20 centimeters (8 inches) on the Waal.

In rural areas, the winter dikes could potentially be set back on a large scale. Some 40 locations were evaluated for large-scale dike setback. The predicted effect of this process ranged from a flood elevation reduction of a couple of centimeters to a half meter.

Urban bottlenecks present a bigger challenge. It is impossible to set back the dikes in locations where the rivers pass through some of the cities. The creation of "green rivers," floodways that pass around the outside of the cities, may produce positive results. These rivers are "green" because normally they are used for nature preserves or for recreational or agricultural purposes. When needed to carry floodwaters, they are opened up. Water surface reductions possible with "green rivers" can be as great as 50 to 60 centimeters (20 to 24 inches). The impact of a green river can extend a great distance upstream.

Currently, specific projects associated with these new approaches are just getting started. In addition to these projects in the Netherlands, similar projects are needed in neighboring countries.

Places to Visit

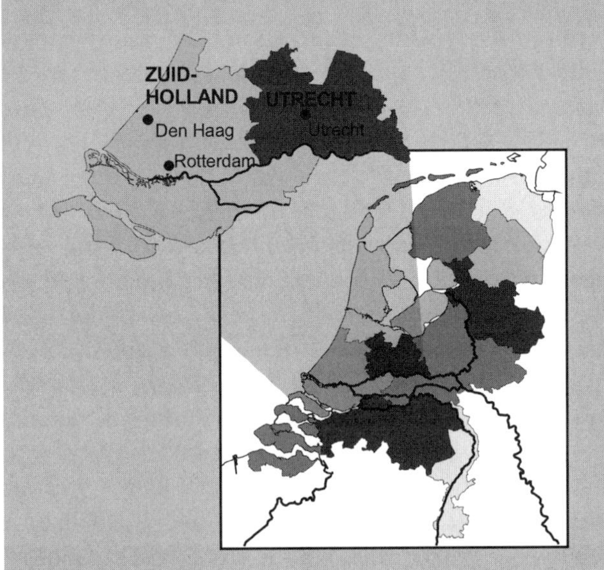

Kinderdijk

DESCRIPTION: This location has more working windmills than any other in the Netherlands. The path running alongside the windmills is always open. One of the mills is open for tours on a regular basis. There is also a small gift shop at the beginning of the path.

LOCATION: Just north of the town of Alblasserdam. Alblasserdam is along A15 about 11 kilometers (6.8 miles) east of Rotterdam in the province of Zuid-Holland.

INTERNET: www.kinderdijk.nl (includes tourist information in English).

EMAIL ADDRESS: info@kinderdijk.nl.

ALONG EXCURSION: Rivers.

Visor Weirs on the Lower Rhine

DESCRIPTION: Weir on the Lek River. This is a semi-circular weir that, when not in use, can be rotated out of the water, allowing ships to pass underneath.

LOCATION: Just east of the town of Vianen, which is located at the A2-A27 interchange in the province of Utrecht.

ALONG EXCURSION: Rivers.

7 Sea Defenses and Coastal Reclamation

Chapters 2 through 6 covered issues of land reclamation and flood protection primarily in inland areas. Chapters 7 and 8 address the issue of flood protection and land reclamation along the coast. This chapter covers the history and technology of sea defenses, with an emphasis on sea dikes. It also covers coastal land reclamation methods. Chapter 8 presents the modifications made to the southwest delta region after a major flood devastated the region in 1953.

The coastal dunes are an important weapon in the fight against flooding from the sea. Unfortunately, the dunes do not protect every part of the coast. In the north, for example, the dunes on the seaward side of the barrier islands are ineffective at protecting the mainland from flooding. Where there are gaps in the dunes, strong sea dikes have been built to protect the mainland. The existing dunes are carefully protected and maintained as an effective barrier.

Protecting the Coast—Dunes and Sea Walls

Sand dunes can be found in most locations along the Dutch coast. (The process of dune formation was described in Chapter 1.) In the southwest the dunes run along the west side of the islands that are formed by the delta. In the central part of the country the dunes form a nearly continuous barrier from Hoek van Holland at the mouth of the Nieuwe Waterweg all the way to Den Helder at the northern end of Noord-Holland. In the North the dunes are present along the west edge of the Frisian Islands (see map in **Figure 1-5**). The outer coastline of the Netherlands, including the northern barrier islands, is approximately 350 kilometers (220 miles) long. Seventy-five percent of this outer coastline, or about 260 kilometers (160 miles), is protected by dunes. Large coastal dikes or sea walls have been constructed along most of the inner coastline and at gaps in the outer coastline (Meijer 1999).

Development of Coastal Sea Dikes

Dikes are earthen embankments used to keep water in its place. The earliest dikes were simply lineal mounds of tamped earth. Because of the erosive action of waves, dikes along the coast were constructed differently than dikes along the rivers or those surrounding drained lake polders. The two important ingredients of a good river dike are the clay lining used to reduce seepage during extended periods of high river levels and the ditch on the landward side used to collect the water that does seep through. The primary requirement of a sea dike is a seaward face that can withstand the wave forces and tidal currents. Methods of sea dike construction varied from region to region.

Up to the thirteenth century sea dikes were not much different from the rest of the dikes. In the thirteenth century a new type of dike was introduced in Noord-Holland. The *slikkerdijk*, or mud dike, was

▸ **Figure 7-1:** Early sea dike cross sections.

Redrawn from Lambert 1971, Fig. 65.

▾ **Figure 7-2:** Dike profiles along the Zuid-Beijerland coast in the southwest delta region. Note: 1 rod equals 5.03 meters (16.5 feet).

Reprinted, by permission, from van de Ven 2004.

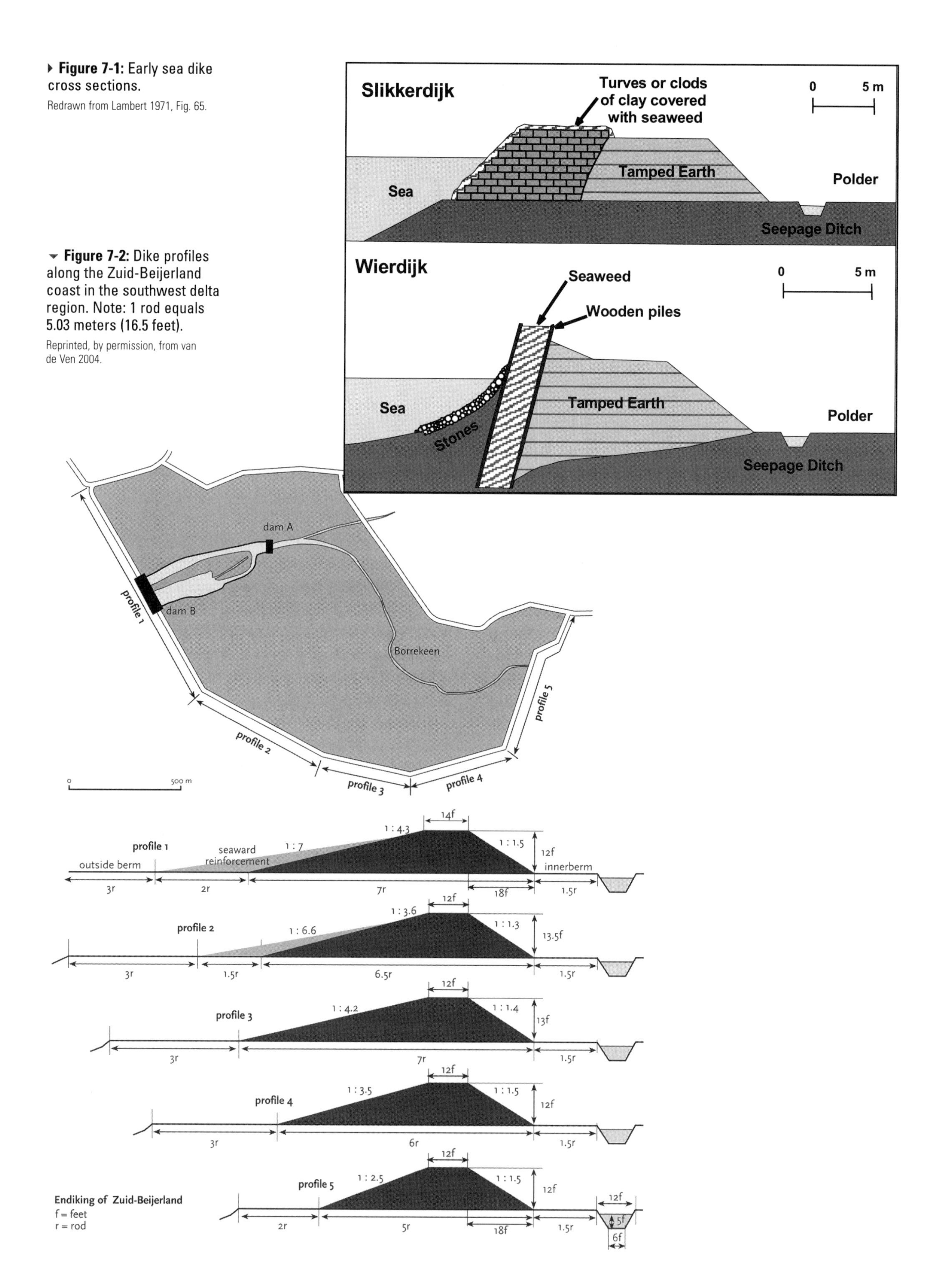

constructed of turves (pieces of sod) or clods of sticky clay laid on the dike's steep seaward side. Not long after construction this side of the dike was covered with a protective layer of seaweed (see **Figure 7-1**). This form of sea dike was used well into the seventeenth century (Lambert 1971).

In the fifteenth century a modification of the slikkerdijk developed. The *wierdijk*, or weir dike, added a wall constructed of seaweed sandwiched between wooden piles. Stone was also added at the foot of the dike. The wierdijk still featured a steep seaward face (see **Figure 7-1**). The seaweed between the wooden piles formed a cushion against the beating of the surf. These dikes reached heights of 5 meters (16.4 feet). The *rietdijk* was a variation of the wierdijk that used a mattress of reeds instead of seaweed. The Diemerdijk, built in 1440 between Amsterdam and Muiden along the Zuiderzee coast, was the first example of this type of dike (Lambert 1971). The pile dike, or *paaldijk*, was a variation of the wierdijk in which the wall consisted solely of 30-centimeter (12-inch) square and 5- to 6-meter (16.4- to 19.7-foot) long wooden piles. This type of dike was used in many places along the shore of the Zuiderzee (including the portion that is now called the Waddenzee along the Friesland and Groningen coastline). It was also used extensively in the southwest delta region and along the lower reaches of the rivers.

In the southwest delta region many of the early dikes that experienced direct attack from the sea were built differently than in the north. They used a cross section with a gentle slope on the seaward side to dissipate the wave and tidal energies. These dikes, called "Zeeland Dikes," were constructed following well-established procedures. Contracts were given for 80-meter (260-foot) sections of the dike (van de Ven 2004). Each contractor accepted work on only one section. The contracts specified that the dike construction be done between the end of February and June 24.

To construct these dikes, soil was dug from pits located in the area outside the dike itself. Rows of planks were used to support the wheelbarrows that hauled the dike material from the pits to the construction site. The dike was built in layers, and horses compacted the soil of each layer. Specifications called for a 2-meter (6.6-foot) overlap of layers between neighboring contractors. Once the soil was in place and inspected, the dike was covered with clay sods 16 centimeters (6 inches) thick. If no clay sods were available, fascines made from rush or straw were used. Because of expected settlement, the dikes were constructed 34 centimeters (1 foot) higher than any neighboring dike.

Figure 7-2 shows typical cross sections used. These were used on the island of Zuid-Beijerland in the southwest delta (part of Beijerlanden as seen in **Figure 1-12**). The areas that received the most direct wave attack were designed with the gentlest slopes.

Sea dike maintenance in the southwest delta area presented some unique challenges. The submerged banks on which the dikes were built consisted of thick layers of loosely packed fine sand of a uniform grain size. The underwater slopes were unstable and often failed—especially under the added weight of the dike. When this happened, the sea dike on top of the bank was often damaged or dragged along with the sliding bank.

To fix this problem, the unstable banks were weighed down with rock, rubble, and chunks of clay supported by fascine mattresses made of osier (similar to that shown in **Figure 5-9**). This was done as early as 1664 in areas north of Amsterdam and 1740 in the southwest delta region. In the southwest, these mattresses were constructed nearby in a location close to the low tide line. Once completed, they were floated on currents to the spot where they were to be sunk. When the mattress was at the desired location, it was held between two boats loaded with blocks of cut clay and rubble. The clay and rubble were placed uniformly on the mattress until it sank. It is thought that without the use of these sunken mattresses many of the Zeeland islands would not exist today (van de Ven 2004).

The pile dikes (similar to the wierdijk shown in **Figure 7-1**) used in the rest of the country proved to be quite successful until about 1730 when the region experienced an infestation of the pileworm or *Teredo navalis*. This is a bivalve mollusk with a rasplike mouth and a soft wormlike body that can reach a length of 0.6 meter (2 feet) and a diameter of 0.8 centimeter (0.3 inch). The pileworm bored holes in the exposed wood of sea dikes in order to protect its soft body. The pileworm, also

Figure 7-3: Nineteenth century sea wall cross section.

Reprinted, by permission, from van de Ven 2004.

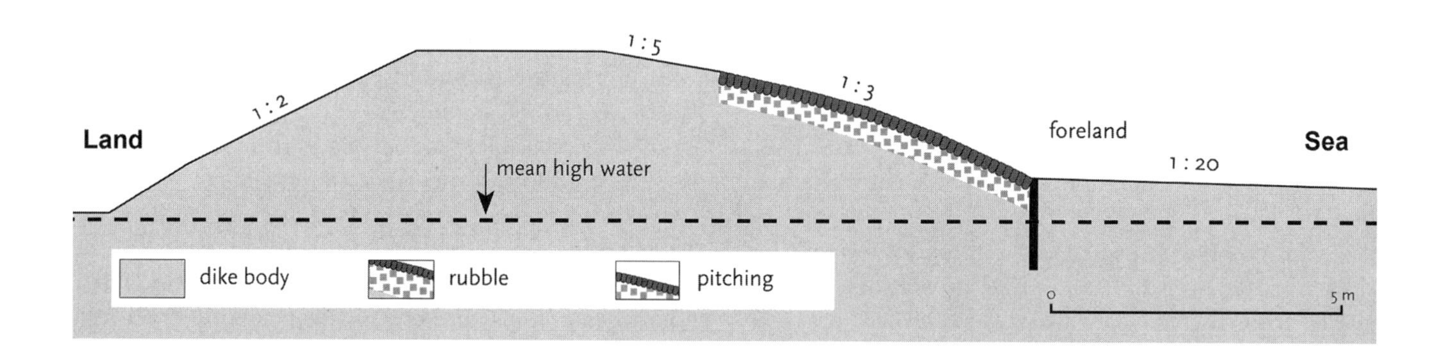

known as the shipworm, had been known to cause damage to wooden vessels as early as the time of the Roman Empire. In the late eighteenth century the British Navy attached copper sheathing to the hulls of their boats, in part, to protect them from shipworm damage. This pest was probably brought to the Netherlands on a wooden ship returning from the East Indies. The invasion of the pileworm happened very quickly. Within a short time, the wooden piles began to crumble, creating a very serious threat to the safety of the country. In 1732, nearly 50 kilometers (31 miles) of the West-Friesland sea dike were destroyed and 20 kilometers (12.4 miles) seriously weakened (Lambert 1971).

Many attempts were made to keep this invasion at bay. These attempts included strengthening the backup sleeper dikes, replacing the softwood piles with tropical hardwood, applying an arsenic compound to the piles, and lining the dikes with iron plates and nails. None of these attempts were very successful; instead, a different approach to sea dike construction was needed. The pileworm problem was finally solved when dikes fashioned after the Zeeland Dike replaced all of the old pile dikes.

By the nineteenth century, sea dike design had advanced to the point where textbooks were being written on the subject. A typical cross section of that era is shown in **Figure 7-3**. A foreland of 300 to 375 meters (980 to 1,230 feet) was recommended for wave run-up. The recommended crest height was 1 to 1.75 meters (3.3 to 5.7 feet) above the highest known storm surge. The crest width was recommended to be 3 meters (9.8 feet) without a road and 5 to 6 meters (16 to 20 feet) with a road (van de Ven 2004). Since safe transportation routes were needed during floods, the road became an important part of the design. If the dike supported a road, it was constructed higher to account for future traffic-induced compaction. The dike body was made of clay delivered to the site by a horse-drawn cart, placed in 0.35- to 0.40-meter (1.1- to 1.3-foot) lifts, and compacted by horses. The slope exposed to wave attack was protected by a layer of stone (referred to in **Figure 7-3** as pitching). The foundation layers beneath the stone included a straw base, followed by brick, and topped with rock rubble. Several different types of stone were used as revetments to protect the seaward face of the dike. The best stone facing was dense basalt columns imported from the volcanic formations of the Eifel region in Germany. When mined, this basalt separated naturally into columns with a uniform hexagonal cross section (see **Figure 5-10**). These columns were then set in place like paving stones, providing a very tight fit. Timber piles were then placed into the dike at the toe of the slope to keep the protective revetments from sliding.

Twentieth century events caused many changes to the standard sea dike design. Many sea dikes were modified in response to storm events in 1916 and 1953. In particular, many dikes were raised and strengthened after these two events. Instead of relying on clay as the primary material for dike construction, sand became the material of choice. Sand was abundantly available from the sea. It was collected using suction dredges and delivered to the site through pipelines in a water slurry. On site, bulldozers pushed the sand over the clay core material to create the dike body. Asphalt, applied in 0.5- to 1.0-meter (1.6- to 3.3-foot) thicknesses, was used as a protective, waterproof outer surface. Concrete blocks were used to further protect the dike face. Construction of new dikes in the southwest delta region will be covered in more detail in Chapter 8.

Andries Vierlingh

Andries Vierlingh lived between 1507 and 1578. As dikemaster to William of Orange, he was actively involved in the construction of sea dikes, especially in the southwest delta region. He acquired a great deal of knowledge about the tides and the behavior of the sea. In his position as dikemaster, he once wrote that his colleagues knew as much about the art of dike building as "a sow knows about eating with a spoon" (Spier 1969). In 1570 he wrote a book about dike building and hydraulic engineering. Unfortunately, the manuscript for his book, *Tractaet van Dijckagie*, was lost for 350 years and was not published until 1920.

In his book, Vierlingh describes the methods that were needed to tame the "roaring breaches" (Lambert 1971). He advocated a patient approach, pointing out that "Water will not be compelled by any force or it will return that force unto you." He said that it was necessary to "direct the streams from the shore without vehemence. With subtlety and sweetness you may do much at low cost."

Vierlingh also presented innovations in sluice design and construction. The most notable development was his design of the masonry sluice. In its construction, he advocated removing the weak soils beneath the structure and replacing them with better soils, placed and compacted in 10-centimeter (4-inch) lifts. He proposed sluices designed with two gates instead of one (for redundancy), with vertical sliding gates attached to a windlass to maintain the water at a specific level, and with provisions to seal off the entire structure for maintenance (van de Ven 2004).

Filling the Gaps in the Dunes—The Hondsbosse Zeewering

Prior to much human activity in the northern part of what is now the province of Noord-Holland, the Zijp estuary pierced the coastal dunes and provided drainage for the creeks that ran through the peat bogs (see **Figure 1-6**). This small gap in the dunes was enlarged greatly with the Saint Elisabeth's Day Flood of 1421. In addition to completely destroying the town of Petten, this flood opened a 5.5-kilometer (3.4-mile) gap in the dunes. A series of sea dikes was constructed over the years to fill this gap. The first was built in 1497. In 1745, the sea broke through again, destroying the rebuilt city of Petten (Spier 1969). This time they solved the problem by building a set of three parallel dikes, providing double redundancy. Construction of these dikes began in 1796. The largest dike, called the Hondsbosse Zeewering, ran directly along the coast and was clad with basalt rocks. This dike was referred to as the "Waker" because it was like someone who is always awake and on guard. A secondary dike called the "Dreamer" was built behind the primary dike. A third parallel dike was also built and called the "Sleeper."

Around 1880, the Hondsbosse Zeewering was relocated to its present position. It was raised and further reinforced in the 1970s. It is now 15 meters (49 feet) high and 140 meters (459 feet) wide at the base. A cross section of this sea dike is shown in **Figure 7-4**. The photograph in **Figure 7-5** shows this sea dike today.

Protection and Use of the Dune Environment

The Dutch Dunes are a critical weapon in the defense of the country against the sea. Since much of the land behind the dunes lies below sea level (see **Figure 1-3**), any break in the dunes would have catastrophic results. A number of measures have been taken to ensure the stability and continuity of the dunes. As early as the sixteenth century, the government organized efforts to preserve the dunes by planting dune grass or *marram*. In some locations, backup dikes were constructed behind the dunes. In the sixteenth century these backup dikes became the primary sea defense when the North Sea destroyed the dune on the island of Walcheren.

Between 1611 and 1738, erosion moved the section of beach between Hoek van Holland and Scheveningen 1,200 meters (3,900 feet) inland. In response, wooden pile breakwaters were constructed. By 1800, twelve breakwaters were built. This number

▾ **Figure 7-4:** Cross section of the Hondsbosse Zeewering.
Redrawn from SWAVN 1986, Fig. 29. .

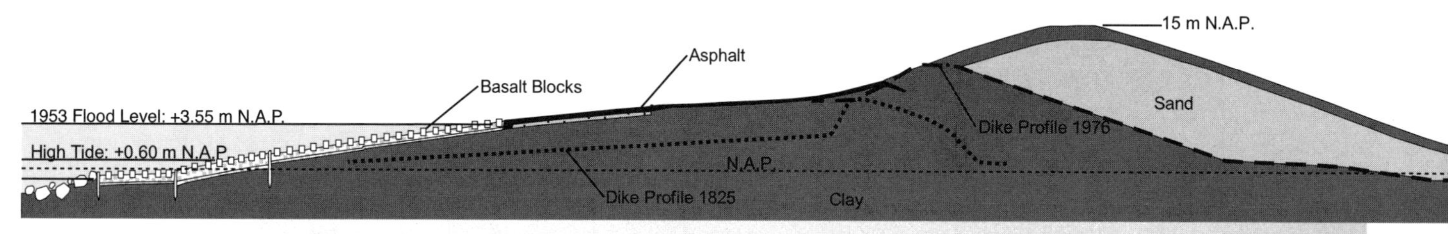

▾ **Figure 7-5:** The Hondsbosse Zeewering looking from Petten to the south.

increased to 40 by the end of the nineteenth century (van de Ven 2004). These still were not completely successful at stopping erosion of the coastline.

Currents and waves are continually moving sand along the coast. In some areas, there is a net loss, resulting in a recession of the dunes. In other areas, there is a net gain. Today, the primary tool used to combat the recession of the dunes is sand replenishment. In this process, sand is dredged from the North Sea as far as 20 kilometers (12.4 miles) from the coast. Trailing suction hopper dredges are used to do this. These dredges are vessels that can dredge and hold up to 20,000 cubic meters (710,000 cubic feet) of sand from depths as great as 100 meters (330 feet). The sand is then transported to the shore, where it is pumped onto the beach. Presently, about 6 million cubic meters (210 million cubic feet) of sand is replenished each year. With continued sea level rise, this amount is expected to increase to 10 million cubic meters (350 million cubic feet) by the year 2050 (CPSL 2001). Sand replenishment locations are determined based on a yearly mapping of the coastline. The objective is to maintain the Dutch coastline at the 1990 reference position (Meijer 1996).

It should be noted that, even at locations that are most at risk, hard measures, such as dams and dikes, are no longer being considered to maintain the coastline. Only soft measures, such as sand replenishment, are applied today. In the long run, hard measures are usually unsatisfactory. For example, the Hondsbosse Zeewering, described earlier, now juts out into the sea because the coast to the north and south has since eroded and moved inland. Today, most of the dune areas are designated as nature reserves. Access is limited except along designated trails.

The dunes are also used in several places for the storage and processing of drinking water. Water, taken from the rivers and transported through long pipelines, is infiltrated into the dunes at higher levels and extracted from the dunes at lower levels. This process is one step in the treatment of the drinking water and produces water of uniform quality. The infiltrated water also helps maintain the groundwater reserves in the dunes and the

▸ **Figure 7-6:** East-west cross section through the low Netherlands showing the condition of the groundwater.
Redrawn from IDG 1994, p. 7.

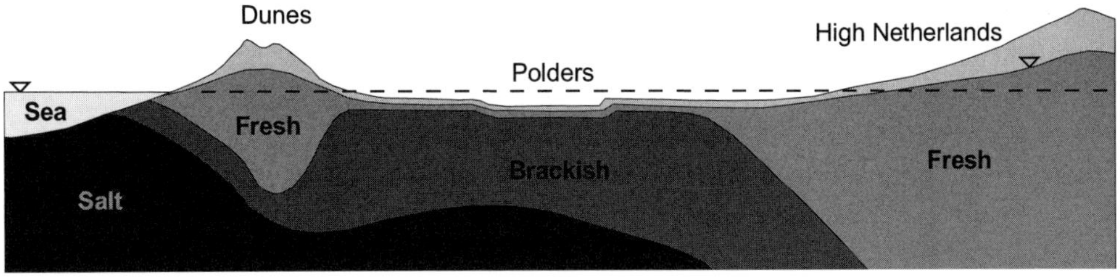

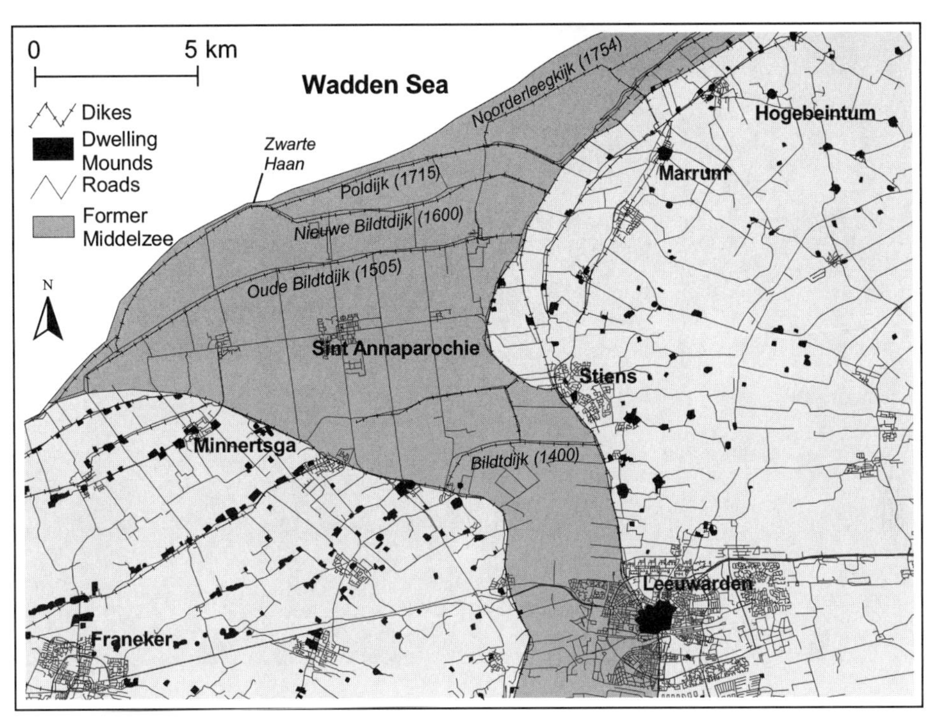

▸ **Figure 7-7:** Reclamation of the former Middelzee.

freshwater bubble along the coast. Since freshwater is lighter than saltwater, the bubble restricts the flow of saltwater into the polders. This is shown in **Figure 7-6**.

Coastal Land Reclamation

Land has been reclaimed along the coast for many centuries. There are both advantages and challenges to coastal reclamation. The obvious challenge is that newly reclaimed land needs to be protected from the onslaught of waves and tides. The advantage is the presence of tides. Reclamations close to the sea can be drained easily down to the level of the low tides. Also, the tide can be used to deliver reclamation fill material. Three different methods for coastal reclamation are described next.

Progressive Dike and Drain Methods

Early coastal reclamation used dike and drain techniques. Following this simple approach, a dike was constructed isolating an area of tidal marsh that had silted up to such an extent that it was infrequently flooded. Drainage ditches were then dug in the endiked area. Water flowed out of the area at low tide through sluice gates that were installed in the dike. The endiked land then drained to a point at which it could be used for agriculture. This method was similar to early methods used to drain inland peat swamps, but here it was applied to shallow coastal estuaries. Eventually, the areas outside the dike silted up to a level at which further reclamation could be done. Several coastal estuaries were reclaimed using this process. The Middelzee in Friesland and the Dollard in Groningen are two examples.

In the Middelzee, sea dikes were constructed along the edge of the estuary and sediment accumulated outside the dike, gradually creating dry land. When it was safe, another parallel sea dike was constructed until the entire estuary was reclaimed. The former extent of the Middelzee can be seen in **Figure 7-7** by following the sequence of parallel dikes leading to the present coastal sea dike. Villages formed behind these dikes. For example, the town of Sint Annaparochie was established

Figure 7-8: Reclamation along the Waddenzee coast.
From AVIODROM Aerial Photography, Lelystad, NL.

behind the Oude Bildtdijk—constructed in 1505. The sea dikes built later include Nieuwe Bildtdijk (1600), Poldijk (1715), and Noorderleegdijk (1754). This area is now lower than its surroundings. Water is currently removed by a pumping station at Zwarte Haan (Black Rooster).

The Dollard estuary was located in the north along the German border. Storms in 1362 first breached the dikes, causing significant expansion. As a result of additional storms and dike breaches, the Dollard reached its greatest extent by the beginning of the sixteenth century. At that time, a dike was constructed strong enough to halt its growth. Finally, dike and drain techniques were used to regain the lost land. The Dollard reclamation is interesting because its newer polders are situated higher than its older polders (Meijer 1996). As areas were reclaimed, the size of the remaining Dollard estuary decreased, resulting in higher tide levels. As a result, sedimentation had to reach higher levels in order for the next area to be reclaimed. The polder elevation differences range from 0.3 to 1 meter (1 to 3.3 feet). As a result, the drainage canals in the newer polders were dug deeper to accept the water from the older polders. Gravity drainage was used in this area until the end of the eighteenth century, when windmills were first introduced.

Using the Tides to Reclaim Land from the Sea

Along the Waddenzee coast a different method was used to gain land. The Waddenzee is the shallow sea that lies between the Frisian Islands and the mainland. Much of the sea bottom is exposed with each low tide. Around 1930 methods were developed to gain land along the Frisian and Groningen coast by encouraging sediment deposition. This method is referred to as the Schleswig-Holstein system (Volmer et al. 2001).

Osier breakwaters were constructed on a 400-meter (1,312-foot) rectangular grid, forming 16-hectare (40-acre) sedimentation fields. Earthen dams divided the sedimentation fields into 1-hectare (2.5-acre) plots. Gaps in the earthen dam allowed tides to penetrate into the area. Small ditches, cut parallel to the coast, drained the plots. When the sediment delivered with each tide accumulated to a sufficient depth, the area was endiked and became good arable land. Around 80 square kilometers

Figure 7-9: Rotterdam-Europoort area showing the Maasvlakte.

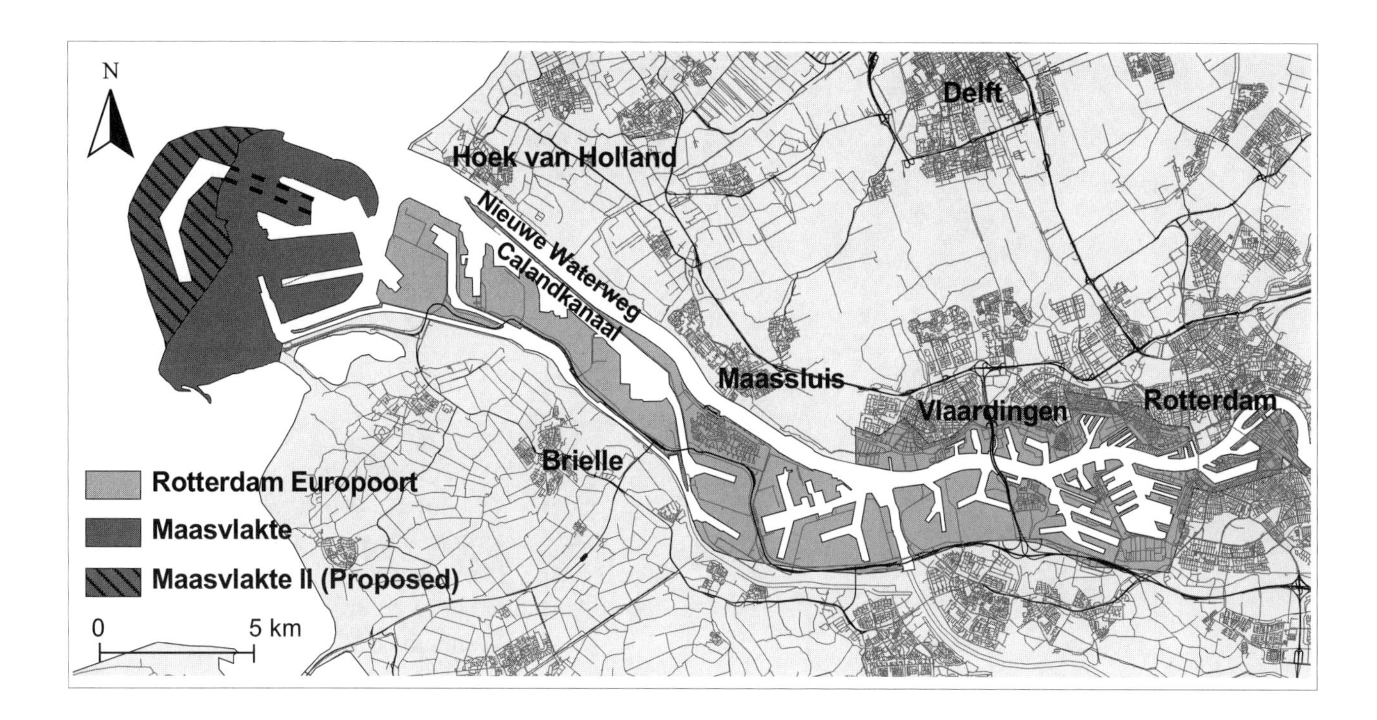

(30 square miles) have been reclaimed in the Netherlands using this method. The photo in Figure 7-8 shows this in progress. Since the Waddenzee is now considered to be a valuable natural area, this type of reclamation is no longer allowed.

Creating New Land from Dredged Soil

Another method used to create new land along the coast is filling shallow areas with dredged sand. The most successful application of this method is the expansion of the Europoort harbor area into the North Sea.

Rotterdam harbor suffered great losses during World War II. After 5 years of war, the harbor area lost 35 percent of the quay walls, 45 percent of the transshipment capacity, and 30 percent of the storage space (Hupkes n.d.). Recovery came quickly. As early as 1952, shipping activity returned to prewar levels. With further growth in shipping, the harbor facilities started to expand in a seaward direction. This expansion included new dock facilities, large petroleum refineries, and facilities to handle container ships. Soon, the Rotterdam-Europoort harbor area expanded all the way to the Dutch coastline. Beginning in 1974, sand that was continually dredged to keep the shipping routes open was used as fill to allow the port facilities to expand beyond the coast.

One-hundred-seventy million cubic meters (6 billion cubic feet) of sand was placed, creating 30 square kilometers (11 square miles) of new land. This land is now used for shipping, materials processing and storage, and warehouses. The western end of the port, called the Maasvlakte, can easily accommodate the largest ships. Figure 7-9 is a map of Rotterdam-Europoort showing the Maasvlakte expansion at the western end. Plans are now in motion to expand the Maasvlakte further. The Maasvlakte II (also shown in Figure 7-9) will expand the port by 10 square kilometers (4 square miles). Construction is anticipated to begin in 2006 (Hupkes n.d.).

Places to Visit

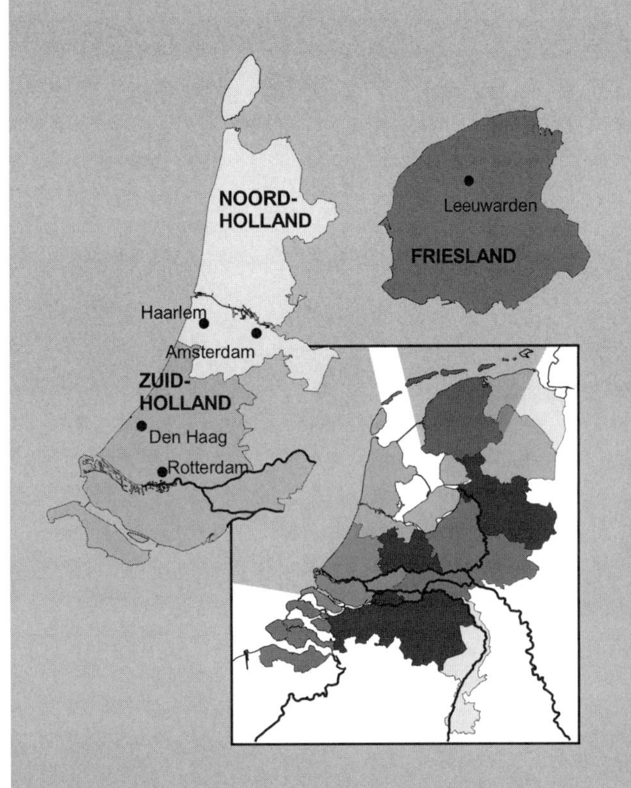

Hondsbosse Zeewering

DESCRIPTION: This is the large sea dike that closes off a major gap in the dunes along the coast of Noord-Holland. There are several viewing points along its length.

LOCATION: The dike runs between the towns of Petten and Camperduin in Noord-Holland. Access to either end can be reached from N9 north of Alkmaar.

ALONG EXCURSION: Noord-Holland Drained Lakes.

Zwarte Haan

DESCRIPTION: This is the location of the pumping station that drains the former Middelzee. The drive to this location crosses several old sea dikes that were built in the process of reclaiming this area. This is also a good location to view the Waddenzee.

LOCATION: On the Wadden Sea coast about 16 kilometers (9.9 miles) northwest of the city of Leeuwarden in the province of Friesland.

ALONG EXCURSION: Frisian Coasts and Dwelling Mounds.

Maasvlakte

DESCRIPTION: This is the 3,000-hectare (7,400-acre) shipping and warehouse space created at the western end of the Rotterdam-Europoort harbor. Several high viewing locations can be found along the dunes.

LOCATION: At the westerly end of A15/N15 approximately 30 kilometers (18.6 miles) west of Rotterdam in the province of Zuid-Holland.

ALONG EXCURSION: None.

8 The Delta Project

In the twentieth century, the Dutch tackled ever-larger land reclamation and flood protection projects. The Zuiderzee reclamation (presented in Chapter 5) was easily the biggest project ever attempted in the Netherlands to regain lost land. The Delta Project was the biggest project ever attempted in the Netherlands to provide protection from North Sea flooding. The stories of these two projects are similar in many ways. In both cases, plans were considered initially without any urgency. In both cases, a flood event put the plans into action. Both projects were tackled in stages, starting with the easiest parts and proceeding to the hardest. The time taken to complete each project exceeded the original estimates. Finally, by the time each project was complete, the needs of the country had changed, resulting in new approaches, configurations, and developments.

Chapter 1 included a discussion of developments along the southwest coast. This area is formed by the deltas of the Rhine, Maas, and Schelde rivers. Over the last 2,000 years, the configuration of the islands and river mouths has changed dramatically, both by natural causes and by human activities. **Figure 1-12** shows the changes that occurred between 1250 and 1600. In 1953, a combination of high spring tides and a storm surge broke many dikes in this part of the country. Vast portions of the delta were flooded and over 1,800 people lost their lives. The Delta Plan implemented as a result of this flood disaster involved damming the tidal estuaries, thereby creating a stronger barrier at the coast. The project included the construction of many dams, several storm surge barriers, locks that separated salt and fresh water, and higher sea dikes. Concerns over the environmental impact of this project resulted in significant design changes and will likely result in future operational changes.

The Southwest before 1953

Figure 8-1 shows a map of the Dutch southwest in the 1940s. Most of the land in this region lay between –2 and +5 meters (–6.6 and +16.5 feet) NAP. The island of Schouwen-Duiveland had the largest areas below sea level. The islands were protected by a system of dikes, some dating back to the twelfth century. The few bridges made the area difficult to access, resulting in poorer economic conditions than in the rest of the country. There were a large number of ferry services, but these were slow and unreliable in poor weather. In 1939 the Storm Surge Committee, appointed by the Minister of Transport and Water Affairs, warned about the poor condition of some of the sea dikes.

The first of two major disasters occurred near the end of World War II. In October 1944, in an effort to bring the German occupation to an end, the Allied forces bombed the dikes at four locations on the island of Walcheren. They were successful in moving the German forces away from the heavily guarded Westerschelde

Figure 8-1: Southwest delta around 1950.

Redrawn from van de Ven 2004, Ch. 10, Fig. 20.

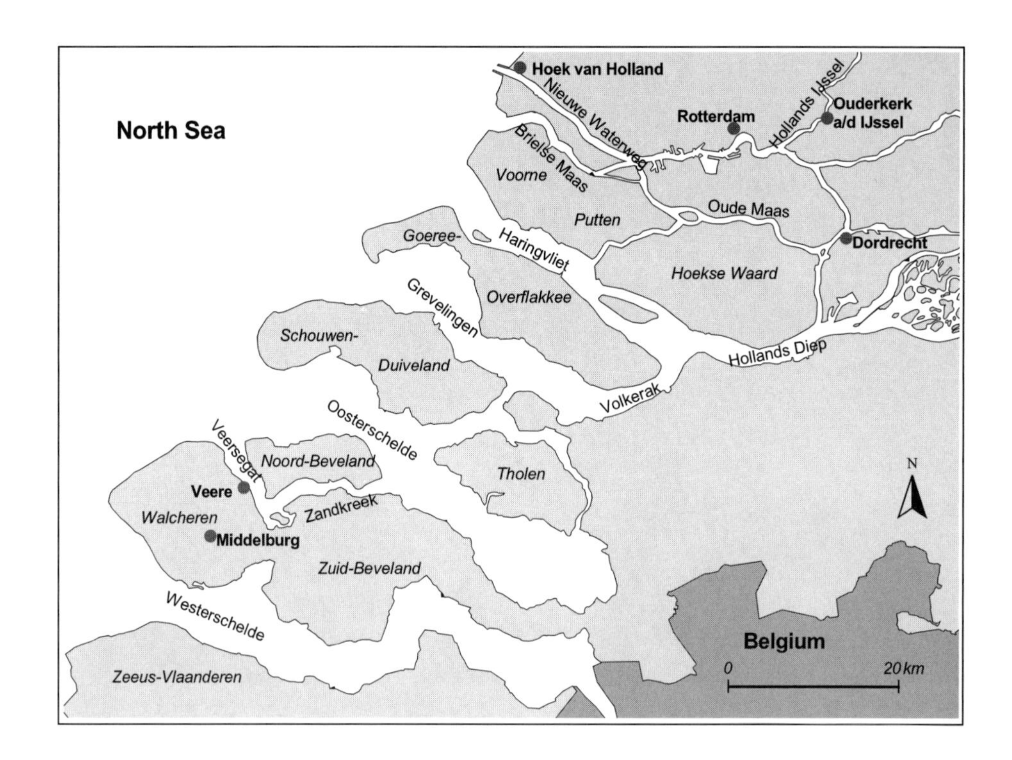

estuary. They were also successful in flooding 160 square kilometers (62 square miles) of land.

In most locations, after a flood has occurred, the standing water simply drains away from the inundated areas, allowing the site to dry out eventually. In areas that are below sea level (both at average and at high tide), the process is much more difficult. The water does not drain away naturally, but continues to flow in and out of the gap in the dike on a daily basis. The longer the dike goes unrepaired, the more damage is done. The soft soils at the dike breach are also eroded away, creating new tidal inlets.

This is what happened on Walcheren. Little could be done to repair the gaps created by the Allied forces until after the country was completely liberated in May 1945. Help came in June 1945 from the contractors that were mobilized around the Zuiderzee reclamation. In 1932, those working on the Zuiderzee reclamation were able to successfully fill the last gap in the construction of the Barrier Dam. It would not be that easy along the coast of Walcheren. Here the tidal variations were three times those in the Zuiderzee. The resulting velocities through the gaps were therefore far greater. It was impossible to close these gaps by simply protecting the sea bottom with fascine mattresses and dumping boulder clay in the gap as was normally done. Furthermore, the boulder clay that was used so successfully in the Zuiderzee project was unavailable in the southwest.

After many months of work, the final gaps were closed using some rather unconventional means. The end of the war left a great deal of naval hardware that was no longer needed. The British had designed and constructed components for the "Mulberry" harbors that were used to off-load massive amounts of supplies directly after the D-Day invasion. These harbor pieces were floated across the English Channel and set in place to create two 3.2-kilometer (2-mile) long harbors. The components included concrete "Beetle" pontoons (to support the pontoon bridge that connected the pier heads to the shore) and larger "Phoenix" caissons (to form the actual breakwater). The largest Phoenix caissons were 69 meters (226 feet) long, 19 meters (62 feet) wide, and 19 meters (62 feet) tall. Unused Beetles, Phoenix caissons, and other surplus ships were available to help close the gaps.

By the time the workers were mobilized to fix the gaps left by the Allied forces, tidal currents had already dug deep channels. The workers first constructed a new dike that formed a half-circle ring, which was connected to the original breeched dike on either side of the hole. The section of this new dike that passed through the shallow areas was constructed first. The tidal channels leading to the breach were filled next, starting with the smallest and continuing to the largest. The bottom of each channel was protected using fascine mattresses to reduce erosion and limit any expansion of the channel as it was filled. The last gap was the hardest to fill because the tidal current velocities were very high. This was eventually accomplished by floating pontoons, caissons, and surplus ships into the gap. At the change in tide they were sunk, thereby filling the gap.

At Rammekens (east of Vlissingen) 750 meters (2,500 feet) of the dike were destroyed. The patch dike around the gap was located 150 meters (490 feet) inland. At this location, five distinct gullies had been eroded, passing 10 to 24 million cubic meters (350 to 850 million cubic feet) of water at each tide (van de Ven 2004). After several attempts, one of which scoured a hole 25 meters (82 feet) deep, this gap was closed with a combination of scuttled ships, Beetle pontoons, Phoenix caissons, and torpedo netting. It was completely closed on February 3, 1946.

When the island of Walcheren was finally pumped dry, the devastation caused by the saltwater flooding became clear. Nearly all plants had died. Mussels grew on dead shrubs and trees. Houses were destroyed, and ditches were silted in with sand.

Changes to the Delta prior to the 1953 Flood Disaster

Prior to 1953, plans were drawn up to make major changes to the southwest coast. These plans were designed to provide protection from both flooding as well as the effects of salt water. In a region in which so much agricultural and horticultural land lies below sea level, salt, which enters primarily from the North Sea, can do as much damage to the land as floodwater. Many agricultural crops require chloride (Cl^{-}) levels below 1,800 milligrams per liter while horticultural crops are more sensitive to salt, with a limit of around 300 milligrams per liter (Lingsma 1966). Milk yields from dairy cattle fall when their drinking water reaches 1,500 milligrams per liter. Compare these chloride limits to the average value in the North Sea of 17,000 milligrams per liter.

Salt enters through a number of routes. The first and most obvious route is seepage under the dunes and dikes. Water is constantly being pumped away from the low-lying polders near the coast, thereby drawing the water table down below that of the sea level. This causes saline water to be drawn from the sea, under the dunes and dikes, and into these low-lying polders. Along the open estuaries and tidal inlets, the saline water can seep through the dikes at high tide. At low tide, the salt tends to remain behind as the water drains. The Rotterdam port area's increase in size after World War II also increased the amount of salt water that could penetrate inland through the open Nieuwe Waterweg canal. Water pumped from low-lying polder areas is often stored in large reservoirs or boezems until it can drain by gravity to the sea. As water evaporates in these boezems, its salinity increases. During the extremely dry summer of 1947, many cows died from drinking this salty boezem water. Large navigation locks can also allow a significant volume of salt into the country.

One way to reduce the amount of salt that enters is to isolate the country from the salt source, then provide a means to flush the salt away. The initial plans for modifications in the Delta region specified damming some of the tidal estuaries. This would keep the salt away and also force higher flows down the other river outlets. These measures were considered necessary to complement plans, already in place, to use weirs to divert more Rhine water to the IJsselmeer (see Chapter 6).

Prior to the 1953 Flood, plans were already in place to dam the Haringvliet and the Brielse Maas (old Maas outlet passing by the town of Brielle), forcing more freshwater down the Nieuwe Waterweg (see **Figure 8-1**). In 1950 two dams were used to close the Brielse Maas, creating a new lake—the Brielse Meer.

The experiences of repairing the damage on Walcheren at the end of World War II provided the basis for the development of new ways to close tidal gaps. This experience led to research on the use of large objects (such as caissons) to close gaps, scour forces, channel bed protection, and the most appropriate foundations to use for enclosing structures. The Brielse Maas was closed using 75 prefabricated caissons, constructed for that purpose. The final gap was closed with two surplus Phoenix caissons (Lingsma 1966). There were two styles of prefabricated caissons—an open style that had to be placed with a crane and a closed style that could be floated in place and sunk.

▾ **Figure 8-2:** Dike burst during the 1953 flood.
From Waterland Neeltje Jans.

◢ **Figure 8-3:** Photos taken after the 1953 flood showing the dike's inner slope failure.
© Ed van Wijk / Nederlands fotomuseum, Rotterdam.

⯯ **Figure 8-4:** Extent of flooding in 1953.
Redrawn from IDG 1996, p. 19.

The 1953 Flood

The year 1953 was a turning point in Dutch flood protection efforts. Only seven years earlier, gaps were being closed in the dikes of Walcheren. Concerns over flood protection and saltwater intrusion started a process of dam construction in the Delta area. Postwar reconstruction efforts were leading to vibrant economic conditions. The Port of Rotterdam was returning to prewar activity levels. It was then that the storm of the century hit, creating a storm surge with water levels on the North Sea higher than ever previously measured.

At 11:00 a.m. on Saturday, January 31, 1953, the Royal Netherlands Meteorological Institute issued warnings that high water levels were likely to occur along the coast. By 6:00 p.m. that day, they suspended pilotage services in the Rotterdam Harbor because of bad weather. Many ships sent distress signals as the evening progressed. Shortly after midnight, the dikes started to break. The toll to the land included 800 kilometers (500 miles) of dike severely damaged, 67 flow gaps (where water is continuously flowing through the gap even at low tide) created, 2,000 square kilometers (770 square miles) of land inundated, and 3,000 houses destroyed (43,000 damaged). The human toll included 1,835 people dead and 72,000 requiring evacuation (van de Ven 2004). **Figures 8-2** and **8-3** show the damage this

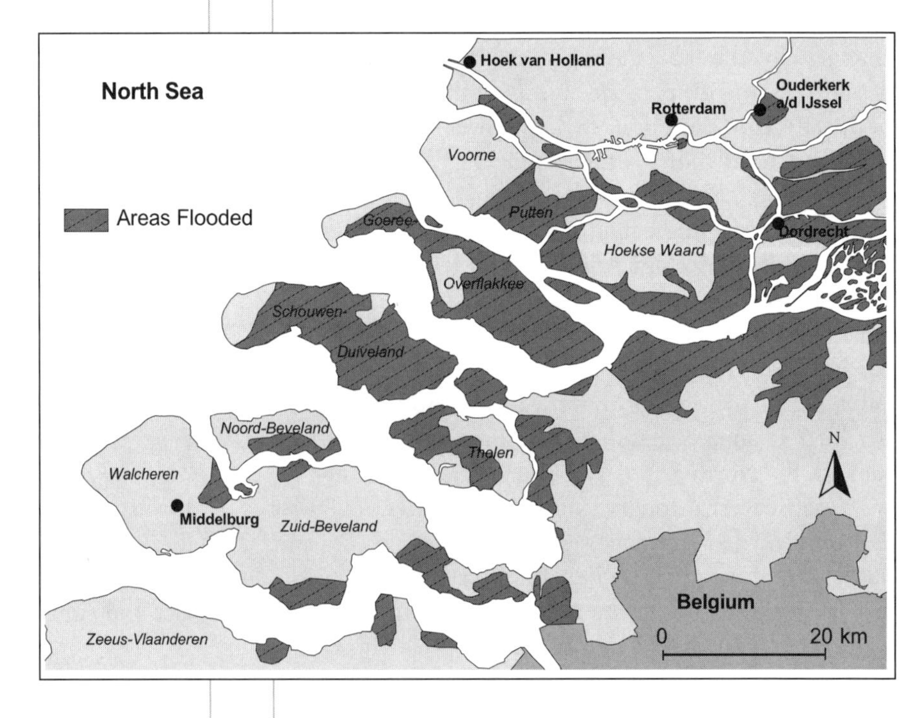

▸ **Figure 8-5:** Gap closure at Schelphoek using an auxiliary dike.

From AVIODROM Aerial Photography, Lelystad, NL.

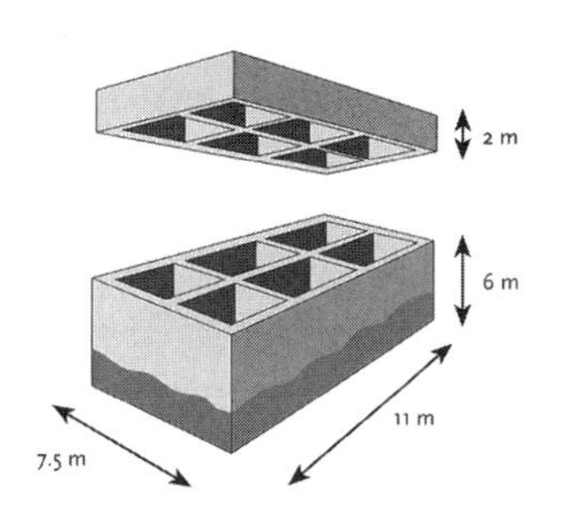

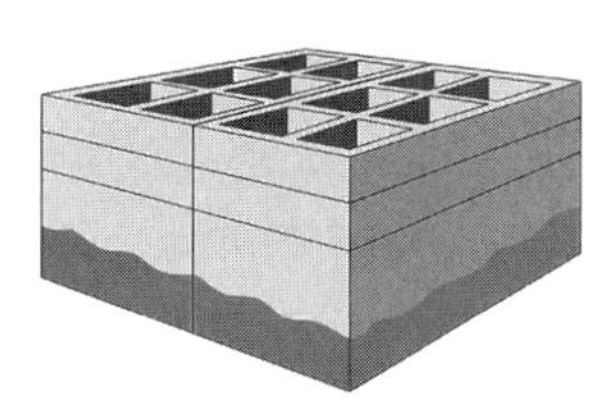

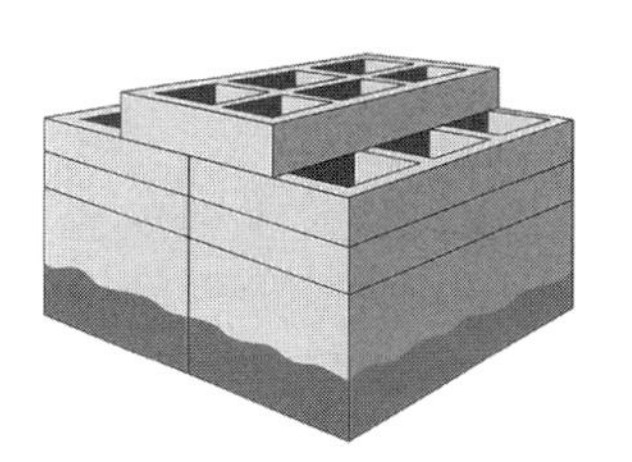

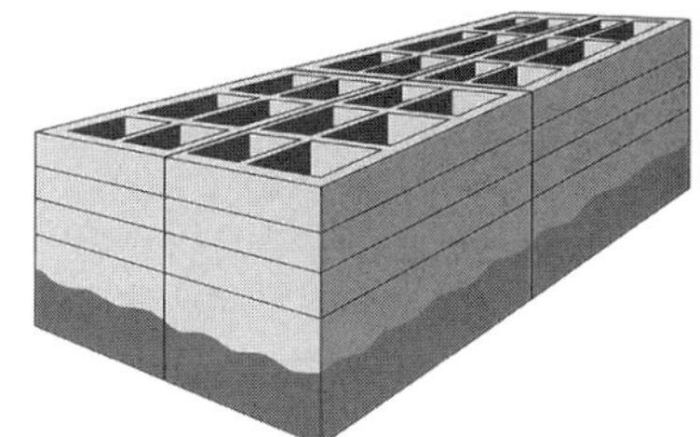

▾ **Figure 8-6:** Standard caissons used for closing tidal channels.

Reprinted, by permission, from van de Ven 2004.

storm caused to two of the dikes. The extent of the flooding is shown in the map in **Figure 8-4**.

A combination of high wind force, unusual wind direction, ill-timed spring tide, and long duration made this storm particularly bad. When the storm approached the Dutch coast, the wind forces were at a level of 10 on the Beaufort Wind Strength Scale. A wind force of 10 is called a whole gale, with wind velocities in the range of 89 to 102 kilometers per hour (55 to 63 miles per hour). The wind direction was also unusual. Normally storms of this type follow a path that results in winds that blow from the southwest—parallel to the Dutch coast. This storm blew from the northwest—perpendicular to the coast—pushing water higher along the coast and into the estuaries. On the night of the storm, there was a spring tide. This occurs when the moon and the sun align to create higher than normal tides. To make matters worse, the storm lasted longer than most—about 18 hours. This gave it time to do significant damage because of long duration of the storm surge. At Hoek van Holland the North Sea water levels were measured at 3.85 meters (12.6 feet) NAP. This is 0.57 meters (1.87 feet) higher than the highest level, previously measured in 1894. This is also 3.05 meters (10.0 feet) higher than the normal high tide water level (van de Ven 2004).

Most of the dikes that failed did so because they were simply not high enough. The revetments, which protect the outer slopes of the dikes from the erosive action of waves, performed well. Yet with such high water levels, the waves were able to flow over the crest of the dikes and scour the inner slope, causing failure (see **Figure 8-3**). In some cases the outside water level was simply higher than the dike crest itself. This was true for many harbor dikes, whose crest heights are lower than that of the sea dikes.

Repairing the Dikes

The Dutch now had to return to the task of repairing the dikes. This time the job was much larger. One advantage after the 1953 Flood was that they were able to start the repair work immediately and use the experience gained from closing the dikes on Walcheren and the Brielse Maas. The government set a goal to close all of the gaps before the next winter. By May 1, 1953, 58 of the 67 flow gaps were sealed. The last nine took much longer. One of the more difficult gaps to close was at Schelphoek on the island of Schouwen-Duiveland. This gap was approximately 400 meters (1,300 feet) long and 40 meters (130 feet) deep (Lingsma 1966). It was closed on August 27, 1953, using a 4-kilometer (2.5-mile) long auxiliary dike including 235 "standard" concrete caissons and one Phoenix caisson. This closure can be seen in the photograph in **Figure 8-5**.

The standard caisson was an adaptation of the prefabricated caisson developed for the damming of the Brielse Maas. Each caisson was 11 meters (36 feet) long, 7.5 meters (25 feet) wide, and 6 meters (20 feet) high. Several are shown in the drawing in **Figure 8-6**. These

Figure 8-7: Closing the gap at Ouwerkerk.
From AVIODROM Aerial Photography, Lelystad, NL.

standard units were constructed off-site and placed together in different combinations to provide the correctly sized closure.

The last gap to be closed was at Ouwerkerk on the south side of the island of Schouwen-Duiveland. This gap was finally closed after the fourth Phoenix caisson was sunk on November 6, 1953. (See photo in **Figure 8-7**.)

The Delta Plan Overview

In response to the disaster brought on by the February 1953 flood, the Minister of Transport and Public Works instituted the Delta Commission on February 21, 1953. This group, consisting mostly of engineers, was assigned the task of finding ways to make sure that a flood disaster of this magnitude would never occur again. At the time there seemed to be two clear options. One option was to raise the level of approximately 700 kilometers (435 miles) of sea walls and dikes throughout the Delta region. The other option was to close most of the tidal inlets and raise only the dikes along the closed coast and the open shipping routes.

Previous studies had proposed closing the tidal inlets for flood protection and salt exclusion. These studies greatly influenced the direction taken by the Delta Commission. The main advantages to this approach were clear: it would limit intrusive construction to the outer coastline, and it would significantly reduce the length of North Sea coast. Shortening the coastline reduces the cost of maintenance and future upgrades. Furthermore, the damming of the Zuiderzee inlet earlier in the century was a major success. It only made sense to continue on the same path.

The Delta Commission was primarily concerned with providing for the safety of the southwest. They did not consider any potential adverse environmental effects of their plan. As the plans pressed forward, they recognized that closure of the inlets would have a significant negative effect on the fishing economy—but the costs were measured in terms of income lost, not in terms of damage to the environment.

By May 1958 both houses of the Dutch Parliament had approved the plan developed by the Delta Commission, named the Delta Act. Prior to the approval of the Delta Act, money was already allocated for construction of the first structure, the Hollandse IJssel barrier.

It was clear that the approved plan, better known as the Delta Project, was of such a magnitude that it would take several decades to complete. This project would also present challenges never before faced by Dutch hydraulic engineers. The Delta Commission set up an order in which the various structures would be built. The first structures were built in locations that were most threatened by another flood. Sequencing was also based on proper development of hydraulic engineering knowledge. They decided to start with smaller, easier barriers and move progressively to the more difficult ones. In this way, lessons learned early could be applied later in the process.

The bulleted list below provides an overview of the actual Delta Project construction sequence. Strengthening of the dikes was done independently of the structures listed below. The original plan had a 1978 completion date. Because of changes made along the way, the last structure was completed in 1997.

The dams and barriers constructed can be classified as either primary or secondary. The primary dams and barriers were located at the mouths of the estuaries (see **Figure 8-8**). These dams separated their respective estuaries from the North Sea. The secondary dams were located farther inland at the tidal divide between two neighboring estuaries (i.e., the location between two neighboring estuaries with no tidal flows). These secondary dams isolated neighboring estuaries from each other. The secondary dams were built first. Once the secondary

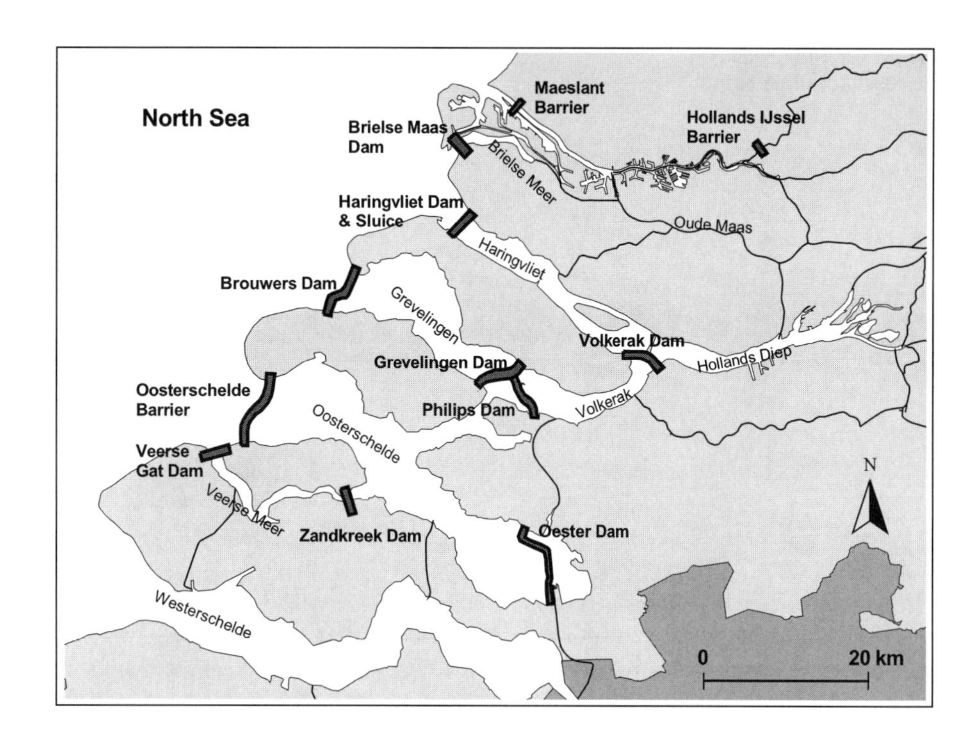

Figure 8-8: Completed structures of the Delta Plan.
Redrawn from IDG 1996, p. 19.

dam was completed, one estuary could be closed off with a primary dam without impacting tidal flows in the other estuary. More detail will be provided later concerning the technologies used to complete some of these structures. The map shown in **Figure 8-8** shows all of these structures (including the Brielse Maas Dam, which was completed before the 1953 Flood).

- **Storm Surge Barrier on the Hollandse IJssel (1954–1958).** This structure protects the low polder land to the north of Rotterdam. It is a structure that is normally open to allow traffic on the Hollandse IJssel River. When floods threaten, it is closed, effectively damming the river for a short period and protecting areas upstream from storm surges on the North Sea.
- **Zandkreek Dam (1957–1960).** This is a secondary dam used to eliminate tidal flows through the Zandkreek channel that separates North Beveland from Walcheren and Zuid Beveland.
- **Veerse Gat Dam (1958–1961).** This is the primary dam, which completed the closing of the Zandkreek channel.
- **Grevelingen Dam (1958–1965).** This was the secondary dam needed to isolate the Grevelingen estuary from the Oosterschelde and the Volkerak.
- **Volkerak Dam (1955–1969).** This secondary dam was used to isolate the Volkerak and Haringvliet estuaries.
- **Haringvliet Dam (1956–1972).** This primary dam closes off the Haringvliet estuary, the primary outlet for Rhine and Maas River water. To allow water and ice to continue to discharge, this dam was constructed with large discharge sluices.
- **The Brouwers Dam (1963–1972).** This primary dam closes the Grevelingen estuary.
- **The Oosterschelde Barrier (1967–1986).** In 1976, midway through construction, the design of this structure was significantly modified because of environmental concerns. To maintain the environment of this estuary, a storm surge barrier was constructed instead of a dam. This barrier allows water from the North Sea to continue to flow in and out with the tides. When a storm threatens, the barrier is closed, protecting the area from floods.
- **Philips Dam (1976–1987).** This dam was constructed after the decision was made to maintain the Oosterschelde as a tidal, saltwater estuary. It created a barrier between the saltwater Oosterschelde and the freshwater Volkerak.
- **Oester Dam (1977–1988).** This was also constructed to provide a freshwater-saltwater barrier. It provided a freshwater, nontidal shipping route between Antwerp and Rotterdam, the Schelde-Rijnkanaal.
- **Maeslant Barrier (1989–1997).** This barrier closed the last possible opening for storm surges in the Rotterdam area. The Delta Plan kept the Nieuwe Waterweg open because of the high volume of shipping

Figure 8-9: The Hollandse IJssel storm surge barrier.

traffic. The Maeslant Barrier is a floating dam that can be used to completely close the Nieuwe Waterweg in the event of a large storm surge.

Prior to 1939, the required dike crest heights were simply based on the recorded water levels from previous storms. In 1939 a government commission abandoned this method in favor of a method based on key factors associated with storm surges. The new methods factored in the probabilities associated with various key events happening simultaneously. These methods also accounted for the impact that proposed delta estuary closures would have on tide levels. The crest heights for the Delta dams were based on storm event water levels with a recurrence interval of one in 10,000 years for economically vulnerable areas and 1 in 4,000 years for other areas (Meijer 1993).

Construction of the Hollandse IJssel Barrier

The Delta Commission identified the Hollandse IJssel River, east of Rotterdam, to be one of the locations most critically in need of flood protection (see **Figure 8-1**). This river is in direct communication with the North Sea via the Nieuwe Waterweg. It flows through some of the lowest polder areas in the highly populated region between Rotterdam and Amsterdam. The dikes along this river were damaged during the 1953 Flood. A major disaster was averted near the town of Ouderkerk when a ship was purposely sunk in a developing dike breach. Had this dike completely given way, a large part of the urban area between Rotterdam and Amsterdam would have been flooded.

The completed Hollandse IJssel barrier consists of two movable gates. Each gate is 80 meters (262 feet) long and 11.5 meters (38 feet) high (Lingsma 1966). The steel gates are strengthened against the force of the water by a 14.8-meter (49-foot) deep arch structure. The normal position of the gates is 12 meters (39 feet) above NAP, allowing ships to pass easily underneath (see photograph in **Figure 8-9**). Each gate is held in place by two 45-meter (148-foot) tall towers. The gates are raised and lowered by means of cables attached to counterweights. Four motors are located in each tower to help raise and lower the gates. When the gate is above the water, two 19-kilowatt (25-horsepower) motors drive it. When the gate is moving through the water, it is driven by two 6.2-kilowatt (8.25-horsepower) motors operating at a lower speed. When the gates are closed, ships can still pass through a shipping lock constructed next to the barrier.

Zandkreek and Veerse Gat

Plans to build two dams, effectively turning three islands—Walcheren, Zuid Beveland, and Noord Beveland—into one, were in place prior to the 1953 Flood. The original purpose of this plan was land reclamation. After the 1953 Flood, the Delta Plan made this project a reality with safety as its primary aim.

This project was started before many of the others for several reasons. Connecting the islands of Noord Beveland to Walcheren and Zuid Beveland (see **Figure 8-1**) would shorten the coastline by 48 kilometers (30 miles) (Lingsma 1966). This location was also good for testing closure methods that might be used on larger estuaries. This project also increased road connections, created a new freshwater recreational lake, and reduced the ferry crossing time from Schouwen-Duiveland by 20 minutes. The major losses caused by this project were the elimination of the shrimping fleet from the town of Veere and

▾ **Figure 8-10:** Completed cross section of Veerse Gat Dam.
Redrawn from Spier 1969, p. 177.

▸ **Figure 8-11:** The last culvert or sluice caisson is pushed into place for the Veerse Gat closure.
From Waterland Neeltje Jans.

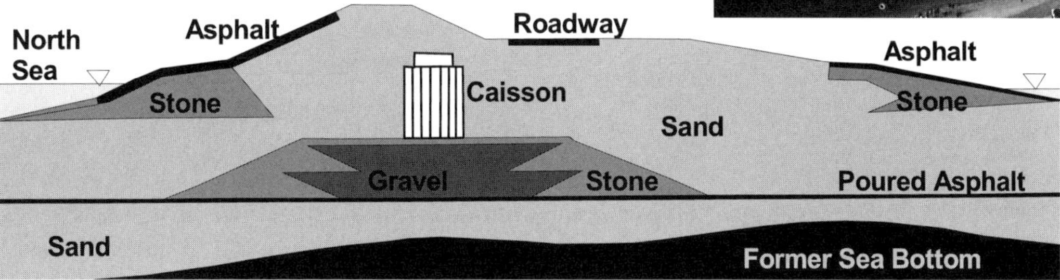

the elimination of the mussel cultivation in Zandkreek. Also, the polders that bordered the new freshwater lake were forced to build pumps to discharge their excess water, since they could no longer rely on low tide for drainage.

The first of these two dams to be completed was the Zandkreek Dam (see **Figure 8-8**). It had to be completed first because it was located at the tidal divide. The final gap in this dam was completed using standard caissons for the dam and the abutment walls. One advantage of standard caissons was that they were stackable, allowing closures with varying depths. Once the closure was made with the caissons, dredged sand was pumped on either side of the caisson wall to complete the dam.

The Veerse Gat Dam (see **Figure 8-8**) presented a bigger challenge. This location was exposed to direct wave attack from the North Sea. It also experienced tidal flows with a volume of 70 million cubic meters (2.5 billion cubic feet) (Lingsma 1966). The implementation of the sluice or culvert caisson made this closure possible. The sluice caisson was patented in 1922 but never used until the closure of the Braakman estuary in 1952. (The Braakman, located along the southern side of the Westerschelde, was dammed prior to the 1953 Flood.)

A sluice caisson was a caisson equipped with removable sides. These caissons were 45 meters (148 feet) long, 20.5 meters (67 feet) wide, and 20.5 meters (67 feet) high (Lingsma 1966). One sidewall of the caisson was built with a temporary wooden wall. The other sidewall had a series of movable steel doors. Each caisson was constructed off-site. With the wooden wall in place and the steel doors closed, it could be floated to the site. Once the caisson was in place, the valves on the bottom were opened and the caisson sunk. The wooden wall was then removed and the steel doors opened to allow tides to flow through. Sand was filled into a ballast box to hold the caisson in place.

Seven caissons were placed over the 300-meter (980-foot) gap without creating excessively high currents. Once all of the caissons were in place, the steel doors were closed at the change in tide. This effectively eliminated the tidal flows on a single day. After they were closed, the caissons were filled and covered with sand, forming the rest of the dam body. The finished cross section is shown in **Figure 8-10. Figure 8-11** shows the placement of the last sluice caisson.

The foundation supporting the sluice caissons was constructed differently at Veerse Gat than at other locations that used caissons for closure. The foundation had to support the caisson and withstand high current velocities due to the reduced flow area. Instead of using the traditional fascine mattress covered with stone, a filter was used. The filter was constructed by placing progressively coarser layers of gravel and rock on the sea bottom. The topmost layer of riprap was heavy enough to remain stable throughout the closure process (van de Ven 2004).

Figure 8-12: Completion of the Grevelingen dam with a cable suspended rock drop.
© Aart Klein / Nederlands fotomuseum, Rotterdam.

Construction of the Grevelingen Dam

The Grevelingen Dam (see **Figure 8-8**) is a secondary dam placed near the tidal divide between two neighboring estuaries. It was placed at the end of the Grevelingen estuary, and it isolated the Grevelingen from the Oosterschelde (see **Figure 8-8**). This placement, at a tidal divide, prevented significant changes in tidal flow patterns from occurring when either primary dam was constructed. The closure was done in two parts to make use of a sand bank in the middle of the estuary. The southern part was closed using standard caissons. Because the northern part was thought to be an easy closure, they decided to experiment with another method—a cable-suspended rock drop.

The cable-suspended rock drop was selected to close the northern section of the Grevelingen because the engineers felt it involved fewer risks than the sluice caisson. First, a sill was constructed to support the dam. This was accomplished by dumping stone on a protective mat. Three different types of mats were used—the familiar fascine mattress, asphalt, and gravel on plastic sheeting. Next, a cableway was constructed over the gap. A cable car was built that ran continuously along this cableway, dropping stones up to 300 kilograms (660 pound mass). These stones were imported from Belgium, Germany, and Scandinavia. Using this method, the closure was built up gradually and uniformly over the width of the gap (see photograph in **Figure 8-12**). When the stone was built up to a level above the water, the rest of the dam was constructed using sand.

Initially, there was some difficulty with cable tensioning. After fixing these initial problems and after sufficient operational testing, round-the-clock dumping operations began in late August 1964 (Lingsma 1966). With eight cars in regular operation, they dumped 12,000 metric tons (13,000 tons) of stone per week. By the end of the year the dam was visible above the water, allowing sand fill operations to start. Operation of the cableway continued until February 9 at a slower pace. Around 2.6 million cubic meters (92 million cubic feet) of sand was pumped behind the stone dam to complete the structure. The dropped stones eventually became part of the slope revetment on the seaward side of the structure. The rest of the dam was protected with asphalt and stone lining.

Construction of Haringvliet Dam and Sluice

The Haringvliet estuary was once the primary outlet for the Rhine River (see **Figure 8-1**). The dam to be built at the mouth of this estuary was the first of three larger primary dams in the Delta Project. Two major challenges had to be tackled at this location. First, during construction this dam had to stop a water volume of 260 million cubic meters (9.2 billion cubic feet) per tide distributed over a 4.5-kilometer (2.8-mile) width. Second, the structure had to allow some Rhine River water and ice to discharge into the North Sea. In fact, this structure was built to be a primary control point for the distribution of Rhine and Maas River water throughout the country. The width of the Haringvliet is 4.5 kilometers (2.8 miles). The closure was constructed as a 3.5-kilometer (2.2-mile) long dam and a 1.0-kilometer (0.6-mile) long discharge sluice complex.

The construction of the discharge sluice complex at Haringvliet presented a major challenge. **Figure 8-13** shows a cross section of one of the sluices. Each sluice bay consisted of two radial gates. These two gates were designed to keep North Sea water from entering the Haringvliet estuary while still allowing excess Rhine water (and ice) to be discharged. The two-gate design was selected for several reasons. First, repair work can be done on one gate while the other is still operational.

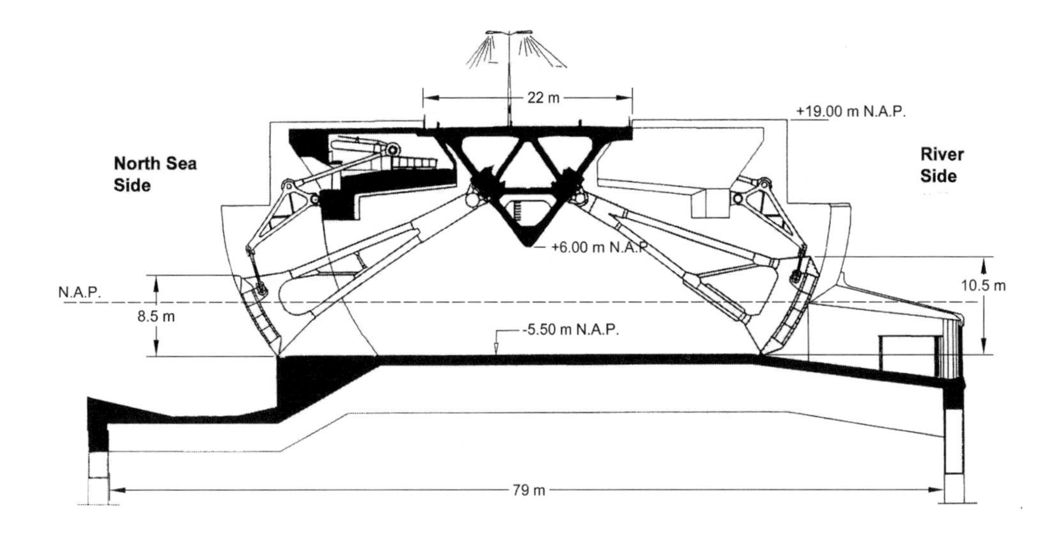

Figure 8-13: Cross section of Haringvliet sluice.

From Rijkswaterstaat, Directie Zuid-Holland, Dienstkring Haringvliet.

Second, the two gates allow for several modes of operation. The sluice is used to control water on both sides. When Rhine River water is discharged, the seaside gates are opened and the riverside gates are used for discharge control. When the North Sea level is higher than the river, no water can be discharged and both gates are closed. When a storm surge occurs, the seaside gates are the primary defense against the waves and high water levels. Since the riverside gates are 2 meters (6.6 feet) higher than those on the seaside, they provide additional protection during extreme storm events.

Early in 1957 construction began on the building pit for the Haringvliet discharge sluice complex (Lingsma 1966). This was designed to be, in essence, a polder in the middle of the Haringvliet estuary. A dike was first constructed all the way around the building site. By November 1957 the dike was almost completed, turning the pool inside the dike into still water. Suction dredgers then entered the pool to excavate the pit down to the desired level. Once they were let out, the last section of dike was finished. The site was dewatered using a system of both shallow and deep well pumps. Pumping began in September 1958. By the end of November the pit was dry enough for construction to start. Since water continually seeped into the site, pumping continued throughout the construction.

In January 1959 they started placing sheet pile walls around the footing of the sluice complex. Driving sheet piles continued for one and a half years. Precast concrete piles from 6.5 to 24 meters (21 to 79 feet) in length were used as foundations for the concrete piers supporting the sluice gates. Between March 1959 and May 1961, 21,840 piles were placed. Trucks passing over the work area on a movable bridge poured the concrete floor of the sluice bays. By October 1961, the concrete floor was in place and construction of the piers was underway.

At the time of its construction, the Haringvliet complex was the largest sluice in the world (to be outdone later by the Oosterschelde Barrier). The 17 sluices were each 56.5 meters (185 feet) wide with a total flow area of 6,000 square meters (65,000 square feet). Each sluice was equipped with two radial gates (see **Figure 8-13**). The gates were supported by 25-meter-long (82-foot-long) steel arms, which pivot on a massive beam constructed between the piers. The 22-meter-wide (72-foot-wide) upper surface of the support beam carried a roadway.

The concrete beam supporting the gates and the roadway was constructed in segments on-site. Each segment was 2 meters (6 feet) thick. The beam section between each pair of piers was constructed out of 22 of these segments held together with 193 prestressing cables. The first beam section was completed by November 1961 and the last in 1964. Installation of the radial gates began in 1963. Each pair of gates took about two months to install.

Once construction of the sluice complex was complete, the dike surrounding the construction site was removed, turning the sluice complex into an island in the middle of the Haringvliet. The rest of the dam body was then constructed. This was done with the sluices left in the open position. The body of the dam was constructed in the same fashion as the Grevelingen, using a cableway. Several improvements were made on the old system. Since this closure was in a location with much higher tidal flows, larger rocks were needed. For this purpose, 1-cubic-meter (35-cubic-foot) blocks of concrete were used. There were several advantages to using concrete blocks. They could be constructed with imbedded hooks, allowing easy lifting and dropping operations. They could also be produced locally. With the dam body finished, the discharge sluices were finally closed, eliminating tidal flows from the Haringvliet estuary. **Figure 8-14** shows the sluice complex under construction. **Figure 8-15**

Figure 8-14: Haringvliet sluice construction pit.
From Rijkswaterstaat, Directie Zuid-Holland, Dienstkring Haringvliet.

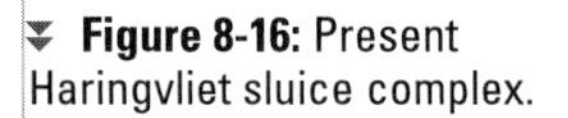

Figure 8-15: Two of the bays of the sluice complex before the gates are installed.
From Rijkswaterstaat, Directie Zuid-Holland, Dienstkring Haringvliet.

Figure 8-16: Present Haringvliet sluice complex.

shows two of the bays prior to the installation of the gates. **Figure 8-16** shows the completed Haringvliet sluice complex.

The Brouwers Dam

The Brouwers Dam was built to cross the 6.5-kilometer (4.0-mile) wide and 30-meter (98-feet) deep Brouwershavense Gat between Goeree and Schouwen-Duiveland (see **Figure 8-8**). There were a number of sand banks in the middle of this gap that allowed for easier construction of the dam. They first pumped dredged sand into the Brouwershavense Gat, combining two of the sand banks into one. This left two gaps to be closed. The gaps had a total length of 3 kilometers (1.9 miles) and a current volume of 365 million cubic meters (12.9 billion cubic feet) per tide. The northern gap was closed with sluice caissons resting on a threshold 8 meters (26 feet) below sea level. The southern gap was closed using a cableway. The cable cars dropped 240,000 concrete blocks, each weighing 15 metric tons (16.5 tons) (Lingsma 1966). **Figure 8-17** shows this dam after completion. The sand that has accumulated along the dam has turned this location into a popular beach area.

The completion of the Brouwers Dam formed a new lake, now called the Grevelingenmeer. The level of the lake was set near the former low tide level at –0.2 meters (–0.66 feet) NAP. At this water level, areas that were previously exposed only at low tide were permanently uncovered. The closure also stopped the tidal flows into this estuary. Because of the sudden change in habitat, many shellfish and plants that needed the tidal fluctuations died within days. Two weeks after the closing, the shores were a huge graveyard of dead animals and plants. A decision was later made to try to repair some of the damage by allowing salt water to enter into the lake from the North Sea. In 1981 a sluice was constructed in the dam to allow this. The Grevelingenmeer has thus become a nontidal, salt lake. Most of the land permanently exposed has now become part of a large nature reserve. With time the lake area has become repopulated with plant and animal species that thrive in this unique environment.

▸ **Figure 8-17:** Completed Brouwers Dam.

From Waterland Neeltje Jans.

◢ **Figure 8-18:** Cutaway view of the Oosterschelde Barrier. (Numbered items are described in text.)

Sketch by Rudolf Das, from Waterland Neeltje Jans.

The Oosterschelde Barrier

The Oosterschelde was the largest estuary to be closed in the Delta Project (see **Figures 8-1** and **8-8**). The gap was 9 kilometers (5.6 miles) wide and the deepest gully was 35 meters (115 feet) deep. The tidal volume was 1.1 billion cubic meters (39 billion cubic feet). The original plan called for a dam, which, like the Brouwers Dam, would completely isolate the estuary from the tidal, salt influence of the North Sea. Work started on this project in 1967. The first step was the creation of three construction islands in the middle of the estuary—Roggenplaat, Neeltje Jans, and Noordland. This left four gaps that had to eventually be closed.

In the early 1970s changes were made in the design of the Oosterschelde Dam. Concerns were raised about the impact that this structure would have on the unique Oosterschelde estuary environment. The clean, salty water of the Oosterschelde provided a rich feeding ground for many types of animals. Mussel and oyster farming took place in this area on a large scale. The mud flats, salt marshes, and sandbars were an important habitat for birds. The book *Limits to Growth*, published in 1972 by the "Club of Rome," was widely accepted in the Netherlands and raised the level of concern. This rising concern about the environment sparked a great national debate about the direction to be taken in the Delta. Should the Delta Plan continue with its philosophy of safety over everything else or should accommodations be made for the environment? This question was answered with a redesign of the entire Oosterschelde Dam. Instead of a structure that created a permanent barrier, an open structure was built. The new barrier design allowed the tides to continue to flow in and out of the Oosterschelde and closed the area off only in the event of a storm at sea. The new barrier design used 62 steel gates, which could be raised or lowered in the event of a large storm. This plan was approved by Parliament in 1979.

Oosterschelde Barrier Overview

Figure 8-18 shows a cutaway sketch of this structure. The foundation consists of compacted sand ①, a bottom mattress and an upper mattress ③, and a block mattress ②. Concrete piers ④ support the sill beam ⑤ and the upper beam ⑦. The gate ⑥ is raised and lowered by operation of the hydraulic cylinders ⑧. When closed, the gate covers the opening framed by the piers, the sill beam, and the upper beam.

▾ **Figure 8-19:** The *Mytilus*—built for compacting sea bottom soils for the Oosterschelde Barrier.
From Waterland Neeltje Jans.

◢ **Figure 8-20:** The *Cardium*—the ship used to lay the foundation mattress for the Oosterschelde Barrier.
From Waterland Neeltje Jans.

Preparation of the Sea Bottom

The first step in the construction of the barrier was to create a foundation for the large concrete piers. Because of the size and depth of the estuary, the piers could not be constructed in place as was done for the Haringvliet discharge sluice complex. Instead, they were constructed in a separate area and delivered to the site by a specially designed boat. The underwater foundation had to be prepared in advance.

First, suction dredgers removed some of the poor quality soil on the seabed. The removed soil was replaced by sand. The entire seabed was then compacted for better bearing capacity by a ship specially constructed for the project. The *Mytilus* (see **Figure 8-19**) carried four large vibrating needles, which could compact the seabed soil to a depth of 18 meters (59 feet) below the sea floor (DOSBOUW 1983). It took this compacting rig three years to complete the seabed compaction.

The seabed was then dredged to create a level surface. Once level, three sets of foundation mattresses were laid. These mattresses were placed on the seabed using another ship built for this project. The mattresses were constructed in a factory and rolled onto a gigantic spool. The *Cardium* (see **Figure 8-20**) carried the spool and placed the mattresses on the seabed.

Each pier rested on three mattress layers. The lower mattress was 200 meters (656 feet) long, 42 meters (138 feet) wide, and 36 centimeters (14 inches) thick. It was essentially a soil filter constructed of layers of fine sand, coarse sand, and gravel. Each layer was separated by a layer of synthetic fabric and reinforced with wire nails. The lower mattresses were placed on the compacted seabed with a 3-meter (9.8-foot) gap between neighboring mattresses (later filled with loose gravel and stone). A second (upper) mattress was placed on top of the first at each pier location. These upper mattresses were the same thickness as the lower mattresses but only 60 meters (197 feet) long by 29 meters (95 feet) wide. A third mattress was placed at each pier location to level the foundation. These mattresses were constructed of concrete blocks varying in thickness from 15 to 60 centimeters (6 to 24 inches) (DOSBOUW 1983).

Concrete Pier Construction

The largest part of the project was the construction of the concrete piers. As mentioned earlier, it would have been impossible to construct the barrier on-site as was done for the Haringvliet sluice complex. Instead, each concrete pier was constructed in a dry construction dock. Once a set of piers was built, the construction area was flooded, and the piers were carried by ship to the barrier site and placed on the prepared mattress foundation. The barrier was constructed in three sections (closing three of the four gaps in the estuary) with a total of 62 gates. This required 65 piers. The piers (and gates) varied in height according to the depth of the channel at the particular location. The piers ranged in height from 30.25 to 38.75 meters (99.2 to 127.1 feet). The largest of the piers weighed 18,000 metric tons (19,800 tons) (van de Ven 2004; DOSBOUW 1983).

Figure 8-21: Concrete pier for the Oosterschelde Barrier under construction.
From Waterland Neeltje Jans.

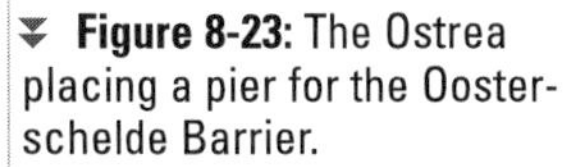

Figure 8-22: Completed piers for the Oosterschelde Barrier wait in the flooded construction pit.
From Waterland Neeltje Jans.

Figure 8-23: The Ostrea placing a pier for the Oosterschelde Barrier.
From Waterland Neeltje Jans.

Each pier consisted of a lower caisson and a tower (see **Figure 8-18**). The caisson was hollow and was filled with sand after being placed on-site. The tower section provided the support for the sill beam, the upper beam, and the roadway. The tower also supported the gate lifting mechanism. **Figure 8-21** shows one of the piers under construction. **Figure 8-22** shows the piers waiting to be floated into position. Each pier took a year and a half to construct, and a new pier was started every two weeks. In the period from March 1979 to January 1983 approximately 450,000 cubic meters (16 million cubic feet) of concrete was processed for the construction of these piers (DOSBOUW 1983).

The *Ostrea* was a ship specially built to carry each pier to its placement site. Once there, it connected with the mooring pontoon, the *Macoma*. The two ships, working together, were able to place the pier within a few centimeters of the desired location (see **Figure 8-23**). The *Macoma* also performed a cleaning process. As the pier was being lowered onto the seabed, the *Macoma* vacuumed the sand off the foundation mattress. After placement, the area between the pier bottom and the foundation mattress was grouted and the caisson section of the pier was pumped full with ballast sand.

Sill Beam, Upper Beam, Roadway, and Gate

The last step was the addition of the various beams and gates between the piers. After the piers were in place, stone was placed between each pair of piers until the desired level was reached (see **Figure 8-18**). Then the sill beam was set in place. The sill beam formed the bottom of the opening through which the tides flow. The sill beams were 39 meters (128 feet) long and 8 meters (26 feet) wide. Stone was then carefully placed on either side of the sill beam, creating a gradual slope from the sea-bottom level to the gate opening.

The last step was installing the upper beam, roadway, hydraulic cylinders, and gates. **Figure 8-24** shows one of the gates being installed. The height of the gates varied to accommodate the variations in sea-bottom depth. The steel gates all had a span of 42 meters (138 feet) and varied in height from 5.9 to 11.9 meters (19 to 39 feet). The gates were constructed of vertical plates on the Oosterschelde side and truss girders on the North Sea side and were designed to withstand loads applied by the

▾ **Figure 8-24:** One of the steel gates being installed in the Oosterschelde Barrier.
From Waterland Neeltje Jans.

▸ **Figure 8-25:** The completed Oosterschelde Barrier.
From Waterland Neeltje Jans.

severest storm (DOSBOUW 1983). Each gate was designed to be raised and lowered with a pair of hydraulic cylinders. The length of these cylinders varied with the height of the gate.

Figure 8-25 shows the completed Oosterschelde Barrier. There were four gaps to be closed between the mainland and the three construction islands. Three gaps were closed using the gated structures described above. The fourth gap was closed with a dam using cableway construction.

Operational Statistics

Figure 8-26 shows the gates in operation during a storm event. In the period between completion in 1986 and January 2004, the Oosterschelde Barrier saw active duty 22 times. All but two closures took place between the months of October and February. Almost a third of the closures occurred in 1990 (seven times). There were eight calendar years during which the gate was never closed. The longest continuous period with no closures was between October 29, 1996, and October 27, 2002 (Ted Sluiter, personal communication).

The Maeslant Barrier

The original 1958 Delta Plan did not plan for any structures in the Westerschelde estuary or the Nieuwe Waterweg. These are both vital shipping routes to Antwerp and Rotterdam, respectively, and cannot be blocked with a barrier. After the barriers laid out in the original Delta Plan were constructed, the Nieuwe Waterweg link remained a concern. The Nieuwe Waterweg connected some of the lowest lying polder land in the country to the North Sea. A ruptured dike during a storm could result in severe flooding throughout Zuid-Holland province. Many kilometers of dikes would have to be raised to provide the needed protection if no structure was built on the Nieuwe Waterweg. Because the Rotterdam-Europoort region was extremely important economically, a high degree of protection was needed.

Another storm surge barrier was proposed for the protection of the entrance of the Nieuwe Waterweg. The engineers' greatest concern was building a structure that would allow ships to pass through the barrier when it was not in use. Ships also needed unrestricted passage during all phases of construction. Several construction companies formed partnerships to get the opportunity to work on this project. These partnerships then presented a number of basic designs. The group BMK (Bouwkombinatie Maeslant Kering) created the winning design and was awarded a design/build contract to construct the Maeslant Barrier (in Dutch, *Maeslantkering*). (The

▾ **Figure 8-26:** Oosterschelde Barrier in use.
From Waterland Neeltje Jans.

▸ **Figure 8-27:** The Maeslant Barrier wall.
From Het Keringhuis.

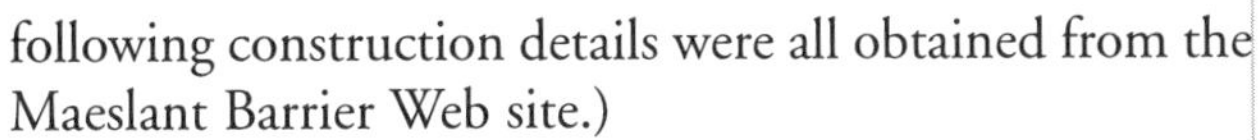

following construction details were all obtained from the Maeslant Barrier Web site.)

The BMK design was based on a pair of hollow, semicircular gates, each attached to pivot points via steel space trusses. The dry docks in which the gates rest are flooded when a storm tide with flood levels 3.00 meters (9.8 feet) above NAP is anticipated. The gates are then floated and pushed into the Nieuwe Waterweg. Once in place, they are filled with water, allowing them to sink onto a prepared concrete sill. This effectively dams off the 360-meter (1,180-foot) wide Nieuwe Waterweg. The photograph in **Figure 8-27** shows one half of the completed barrier. When the storm threat is gone, the gates are pumped dry and repositioned in their docks. The docks are then pumped dry and shipping can resume. At the time of its construction, it was anticipated that this barrier would close, on average, once every 10 years. Between 1997, when it was completed, and January 2004 it has not yet been closed for a flood emergency, only for testing and maintenance.

Construction work on this barrier took place both in the water and on land. In the water, work was done to prepare the sill on which the barrier rests. The sill provided a flat base for the barrier when it was sunk in the Nieuwe Waterweg. The foundation of the sill was designed to remain stable as the barrier was lowered and as the current velocities increased.

The sill was constructed using 64 precast sill blocks. The channel bottom was first prepared for the placement of the blocks. The base foundation was constructed as a porous filter bed. The channel bottom materials were removed to a depth of 8 meters (26 feet) below the existing bottom and replaced with four layers of sand, gravel, and two different sizes of basalt rubble. The total thickness of the filter bed was 2.25 meters (7.4 feet).

The sill blocks were slightly wedge shaped to match the circular shape of the barrier. The blocks were placed within 3.5 centimeters (1.4 inches) of each other. To maintain centimeter accuracy for the placement of the blocks, a 21 meter (69 foot) measuring tower was attached to each block. The measuring towers could be seen poking just above the water surface after the block was sunk into position.

The barrier itself was designed to remain stable under some very difficult environmental situations. Model testing showed that the original barrier design oscillated uncontrollably under certain situations. Design modifications were made to the cross section to avoid this. Most notably, the underside of the seaward side was tapered and stabilizing vanes were added.

The barrier does not completely seal off the flow of water. When closed there is still a 1.5 meter (4.9 feet) gap between the two barriers. If the two barriers were allowed to close completely, they would bang into each other and could easily be damaged.

The barrier wall was built in sections off-site in Krimpen aan de IJssel and in Schiedam. Each wall consisted of 15 upper chambers and 13 lower buoyancy chambers. The sections were delivered by barge and hoisted into the dry dock. Once in the dry dock they were transported on a 24-wheeled truck to the location where they could be welded to the sections already in place. It took five weeks to weld a new section to the existing wall. The wall was designed with additional reinforcing at the six connection points with the trusses. The completed barrier wall was 210 meters (690 feet) long and 22 meters (72 feet) high. **Figure 8-28** shows a cross

▸ **Figure 8-28:** Cross section of the barrier wall.

▾ **Figure 8-29:** Ball joint support of the Maeslant Barrier gate.
From Het Keringhuis.

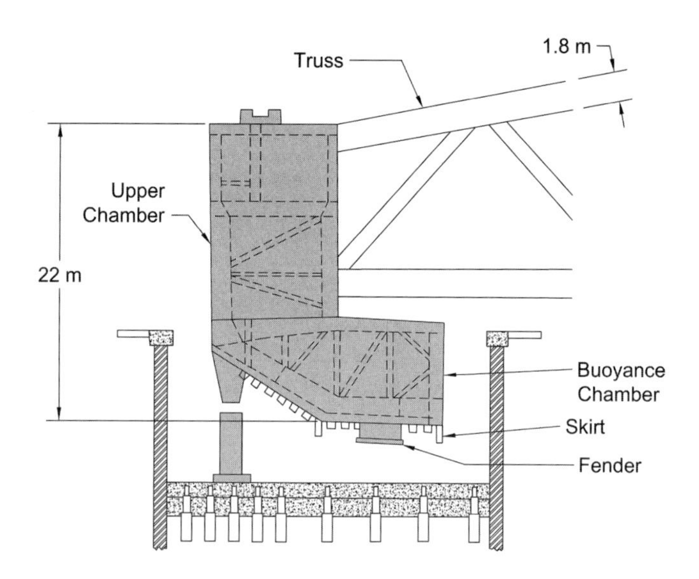

section of the barrier wall. When not in use, the barrier remains inside the dry dock area. This allows for easy maintenance. The barrier is positioned 2.5 meters (8.2 feet) off the floor of the dock and rests on 14 concrete supports.

The barrier wall connects to the pivot point via two trusses. Each truss was 237 meters (780 feet) long. The largest steel tubes in the trusses were 1.8 meters (5.9 feet) in diameter with 90-millimeter (3.5-inch) thick walls. Each welded joint was built of 102 layers. It took about 160 hours to weld each joint.

The pivot was a large ball joint. A ball joint was selected because the barrier wall had to rotate around three axes, and a ball joint provides the freedom to do this. The 10-meter (32.8-foot) diameter ball joints for this structure were three times larger than any previously constructed. Each ball was constructed of a core piece, a front shell, a rear shell, and two lower shells (see **Figure 8-29**). The ball was supported by front, rear, and lower seats. Each component was cast in the Skoda factory in the Czech Republic. **Figure 8-29** also shows the truss connection and the ball foundation.

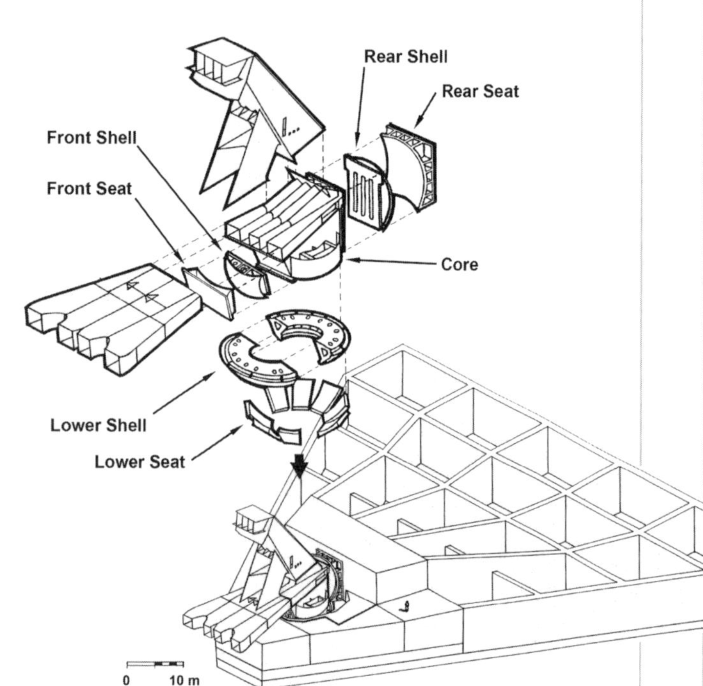

The forces exerted by the water on the barrier wall are transmitted through the trusses and through the ball joint to the ball joint foundation. The foundation was made of a set of concrete cells in a triangular arrangement. The cells were constructed with a ribbed underside to resist base sliding and were filled with soil, giving the entire structure a weight of 520,000 metric tons (570,000 tons). The apex of the triangle points to the middle of the gate when in the closed position. The foundation was designed to carry a lateral load from the barrier dam of 687,000 kilonewtons (154,000 kilopounds). Design calculations predict that, when in use, the foundation could move by as much as 22 centimeters (8.7 inches). Much of that displacement is elastic but some will be permanent.

Figure 8-30 shows the Maeslant Barrier closing the Nieuwe Waterweg entrance to the Port of Rotterdam. This closure was a test of the barrier.

New Plans for the Haringvliet

Prior to 1950 the Haringvliet, the Hollands Diep, and the Biesbosch were part of a tidal estuary—the natural transitional zone between

Figure 8-30: The Maeslant Barrier shown in the closed position effectively blocking the Nieuwe Waterweg shipping canal. The Calandkanaal (Caland Canal) on the right is still open to the North Sea. The port facilities and industries in contact with the Calandkanaal are all built at an elevation that protects them from flooding.
From Het Keringhuis.

the sea and the great rivers. The tidal range varied from 1.8 meters (5.9 feet) at Hellevoetsluis (near the current dam and sluice complex) to 2.25 meters (7.4 feet) at Moerdijk (near the Biesbosch). This was an extensive area of mud flats, creeks, and sand flats populated by a diverse mix of plant and wildlife. The closure of the Haringvliet had a devastating effect on this environment (RWS-DZH n.d.).

The gradual transition from salt water to freshwater was gone. Routes for migratory fish were also cut off. When water was discharged through the Haringvliet sluices, the organisms that could not stand the sudden transition from freshwater to salt water died off. Freshwater fish, flushed out through the sluices, could not return to their original habitat.

The loss of tides increased bank erosion throughout the estuary. With access to tides blocked off, wave action always occurs at the same level. This constant wave action transformed the banks from mildly sloping to steep and eroded. The ground behind the banks dried out and subsided. The banks are now protected with riprap to eliminate further erosion. Furthermore, rare plant communities disappeared and were replaced with more common species. And, since much less water is being discharged, the Hollands Diep and the Haringvliet are now filling up with sediment from the rivers.

In 1994 a study was initiated to look into ways to restore the former Haringvliet estuary (RWS-DZH n.d.). The experience in the Oosterschelde showed that it was possible to maintain some of the former qualities of the estuary by allowing the tides to flow in and out through the barrier structure. In other words, the study was commissioned to ask whether it was possible to operate the Haringvliet Dam as a storm surge barrier, thereby allowing the estuary to recover.

The investigators studied the impact of four different levels of reopening—from fully closed (as designed) to fully open. The study sought to determine the costs and benefits of each level of opening. A fully open barrier maximized the environmental benefits. The primary cost of opening the barrier completely was the relocation of the water intakes that had been built in the 30 years since the Haringvliet was closed. These intakes provided freshwater for municipal and agricultural uses. With the gates fully open, more Rhine and Maas River water will flow through the Haringvliet, leaving less to discharge through the Nieuwe Waterweg. Subsequently, salt water could penetrate up the Nieuwe Waterweg to its confluence with the Hollandse IJssel River, thereby threatening the freshwater supply to the north of Rotterdam. The investigators were also concerned about the hydraulic stability of the Haringvliet discharge sluice itself. It was designed for flows going from the river to the sea and not in reverse. What would be the impact of reverse flows on the structure?

In 2000 the final decision was made to take a controlled tide approach. In this approach, one-third of the gates will be opened 95 percent of the time. This will return a 1-meter (3.3-foot) tidal fluctuation to the Biesbosch, making it again the largest freshwater tidal area in Europe. This approach will require the relocation of several water intakes due to the increased salt levels in the estuary. The current plan is to begin opening the gates in 2008. This will be done gradually—over a period of at least 15 years to allow continual study of the impact of the new operational policy.

Places to Visit

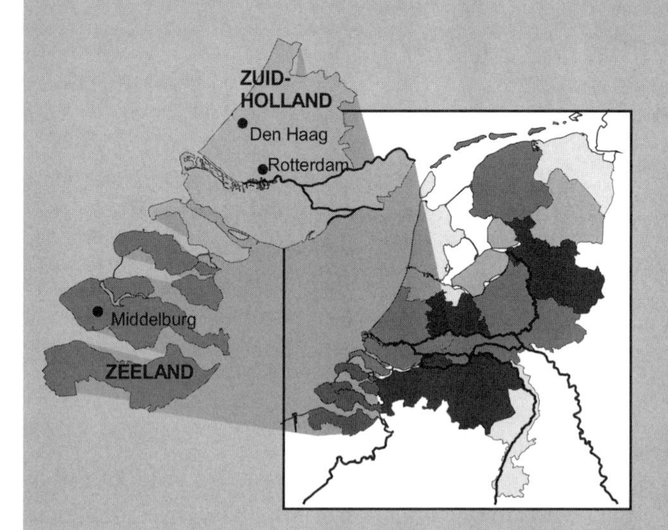

Maeslant Storm Surge Barrier (or Maeslantkering)

DESCRIPTION: This is the storm surge barrier located on the Nieuwe Waterweg. There is a visitor's center providing displays, a view of the barrier, and videos (available in English).

LOCATION: Along the north side of the Nieuwe Waterweg halfway between the towns of Maassluis and Hoek van Holland in the province of Zuid-Holland (west of Rotterdam). Access is from N220.

INTERNET: www.keringhuis.nl (complete version in English).

EMAIL ADDRESS: keringhuis@podium.nl.

ALONG EXCURSION: Delta Works.

Haringvliet Discharge Sluices

DESCRIPTION: This is the location of the discharge sluices for the former Haringvliet estuary. There is a small exposition with a restaurant and tours of the structure.

LOCATION: In the province of Zuid-Holland, at the southern end of the N57 crossing over the Haringvliet sluice complex.

INTERNET: Exposition http://www.expoharingvliet.nl/ (full English version), changing management issues http://www.haringvlietsluizen.nl/.

EMAIL ADDRESS: info@expoharingvliet.nl.

ALONG EXCURSION: Delta Works.

Waterland Neeltje Jans

DESCRIPTION: Waterland Neeltje Jans is the exhibition associated with the Oosterschelde Storm Surge Barrier. This is a combination exposition and water theme park. The exposition highlights are a museum of the Delta Project and a self-guided tour inside the barrier.

LOCATION: In the province of Zeeland, on N57 south of the second set of gates.

INTERNET: http://www.neeltjejans.nl/ (full English version).

EMAIL ADDRESS: info@neeltjejans.nl.

ALONG EXCURSION: Delta Works.

Hollandse IJssel Storm Surge Barrier

DESCRIPTION: This was the first structure completed as part of the Delta Plan. It is a storm barrier used to close off the Hollandse IJssel River. There is no visitor's center or exposition, but there are good places to get a close look at this structure. (See "Rivers" excursion in Chapter 10.)

LOCATION: It is located where N210 crosses the Hollandse IJssel River in the town of Krimpen a/d IJssel in the province of Zuid-Holland. Krimpen a/d IJssel is the first town directly east of Rotterdam.

INTERNET: None.

ALONG EXCURSION: Rivers.

9

Water Management Organization

This chapter covers the history of water management organizations in the Netherlands. In essence, this is the story of the Dutch water boards and of Rijkswaterstaat. The water boards are in charge of local and regional water management. Rijkswaterstaat is the central government agency in charge of water management on a national scale.

History of the Dutch Water Boards

Water management organizations first developed in the early stages of peat bog draining activities. The organizations that developed varied regionally due to variations in political control. In the eleventh and twelfth centuries, the western part of the country was ruled by feudal lords, in particular the counts of Holland and the bishops of Utrecht. These ruling authorities made contracts with groups willing to reclaim a specific area. The contract, called a *cope*, guaranteed that the reclaimers would have full control over the land. In return, they were required to pay taxes to the ruling authorities. Management of land drainage and dike construction remained the responsibility of the local authorities.

The construction and maintenance of a network of ditches and dikes was the responsibility of the village or group of villages. Everyone in the village shared in the responsibility because everyone shared the ditches and dikes. The village council was the main organization in charge of water management. They made agreements with other villages and established the rules of operation and maintenance, called *keuren* (Kaisser 2002). Every farmer was given his share of the duties. The length of dike maintained by a farmer was proportional to the amount of land owned by the farmer. An inspection committee was also established to make sure that everyone did his required job, since the dike was only as strong as its weakest point. In the case of a dike break, the entire community assisted in the reconstruction. Local management was also contingent on finding water discharge routes that did not pass through other communities, something that became more difficult as the land continued to subside.

The weakest part of a community's dike and drainage system was any location where a dike met a stream or canal. Sluices were usually constructed at these points to allow the stream to discharge water through the dike. The sluice was an expensive and complex structure. A sluice could not be operated and maintained in the same way as a dike. Furthermore, the sluice often provided drainage for a large number of villages. To manage this situation, a new administrative structure evolved during the second half of the twelfth century.

The administrative structures that developed were the regional water authorities or water boards, called *waterschappen*. The villages appointed members or *heemraden* to these boards. The authority of the regional water board often extended beyond a dam or sluice; it often took charge of the operation and maintenance of the major dikes and drainage canals in the region. The water boards set up rules of operation and provided inspection of the dikes and dams. They acquired legislative, judicial, and executive powers over issues of regional water management. The villages continued to maintain authority over local issues.

Two of the earliest water boards were established in the twelfth century. The Lekdijk Bovendams water board was established in 1122 when a dam was constructed in the Old Rhine River. The Rijnland water board was established around 1170 by 15 self-governing communities in response to the silting of the Rhine River outlet. This water board was then responsible for finding a new outlet point for local drainage. The Rijnland water board was the first to establish the principle that the amount that anyone pays for drainage works should be proportional to the benefits received. By the thirteenth century most regional water boards had appointed one individual, the *dijkgraaf*, to be the chairman. Being the dijkgraaf or member of the heemraden was a very important position.

Water authorities developed differently in the north. Here the individual peasants had a fair amount of personal freedom. A given peasant was willing to reclaim a portion of land because he was able to fully reap the benefits of his efforts. Regional water authorities developed later, only after the regional administrative structure became more centralized. In the fourteenth century, the monks were the first to start organizing water management on a regional scale. These water authorities involved specific tasks and limited power. For example, the task of a given water authority might simply be the maintenance of a particular sea dike.

Chapter 1 described how human occupation caused the peaty soil to decay and consolidate resulting in land subsidence. As the land subsided, it became more difficult to drain using gravitational forces alone. The development of water-pumping windmills eventually allowed occupation to continue in some of these low-lying polder areas. A polder is a distinct area, usually surrounded by a dike or embankment, within which the groundwater is maintained by a drainage system (usually ditches and canals). In the fifteenth and sixteenth centuries (before lake drainage occurred on a large scale), a number of polders in the western part of the Netherlands were drained by a single windmill. As a result, the polder created a new administrative level in water management and the *polderboard* was born. A polderboard included a heemraden (a board of landowners) and a chairman. This group was in charge of the construction, operation, and maintenance of the polder's windmill (Kaisser 2002).

As a result of the development of the polders in the fifteenth and sixteenth centuries, more water was now discharged into the regional system of storage canals or *boezems*. The regional water boards responded by expanding their control to the administration of water levels and flow rates in the storage canals. This resulted in a two-level system of water management. The polder board managed the water in the polder while the regional water boards managed the water in the storage canals. The members of the regional water boards took on a higher rank and gave themselves a new title—*hoogheemraden*.

As agricultural production improved due to better drainage, the farmers were no longer interested in maintaining their portion of the dike, as had been the case for many centuries. As a result, in the sixteenth and seventeenth centuries, the regional water boards began to use a professional labor force to maintain the dikes. An entire new industry developed led by trained hydraulic engineers. Consequently, the quality of the dikes improved and the frequency of flooding was reduced significantly.

From 1795 to 1813, during the time of French occupation, a national water administration was formed called Rijkswaterstaat. Rijkswaterstaat was given the responsibility of managing projects at the national level. This added a third level to the existing system of water management.

In the nineteenth century several changes were made to the system of water boards. The new constitution of 1848 removed the water board's judicial powers. Additionally, the new constitution brought more uniformity to the entire system.

Two major changes took place in the twentieth century. In 1970 the water boards were given jurisdiction over water quality issues. The number of individual water boards also decreased dramatically. In 1850 there were 3,500 individual water boards, both local and regional. As of December 2004, there are 25 regional (or integrated) water boards.

Today the water boards are responsible for local and regional flood control, water quality, and the treatment of urban wastewater. They manage pumping stations, run wastewater treatment plants, and maintain waterways and flood defense structures.

The water boards levy two taxes to finance their work, the water board charge and the pollution tax. The water board charge is payable by residents, property owners, and landowners. Every person in the Netherlands receives a tax assessment. The water board charge for owners of property and/or land is based on the amount of land they own. The pollution levy is based on the principle that a polluter must pay for the pollution that he or she causes. Every household in the Netherlands pays a pollution tax based on the number of people living in the household. (Households with three or more persons pay the same amount.) Companies and organizations pay an amount based on the quantity and composition of their wastewater.

Life in the low parts of the Netherlands could not exist without careful water management, effectively provided by the local and regional water boards. These institutions have traditionally listened to the concerns of all. The result is a water management system that is based largely on consensus and compromise. This tradition of consensus and compromise is also seen in other political institutions in the country (Kaisser 2002).

Rijkswaterstaat

Until the end of the eighteenth century, all water management activities were in the hands of local or regional authorities. This began to change when, in 1798, a national administration was established to provide technical hydraulic engineering support. This organization, called Rijkswaterstaat, was created at the beginning of the Batavian Republic—the period of French rule in the Netherlands. The establishment of a central authority for public works was primarily a result of political change brought on by a new government rather than a recognized need for centralized water management (Linsten 2002). After the fall of Napoleon, the Netherlands became a kingdom, under the reign of the House of Orange. The new government remained a strong centralized state. A permanent role for centralized water management was firmly established.

Early Rijkswaterstaat engineers directed the defense of river dikes when floods threatened or provided technical direction for the construction of large hydraulic projects. The engineers who first went to work for Rijkswaterstaat had difficulty establishing credibility in an age in which the politicians were members of the aristocracy while the engineers came from a lower class of society—tradesmen, surveyors, carpenters, and so on. The credibility gap narrowed with increased professional training and with project success.

Rijkswaterstaat had several successes in the first half of the nineteenth century. The agency worked to gather and organize hydraulic engineering knowledge from all areas of the country. They established a flood warning system that used engineers, stationed at various locations around the country, to gather and communicate information about water levels, ice dams, and threats of dike bursts. In 1829 they began a process of mapping all of the rivers and establishing gauging stations.

After the southern part of the kingdom broke away in 1830, Rijkswaterstaat saw a period of decline because operating funds were diverted to military operations. Their authority had diminished considerably by the time the new constitution was adopted in 1848.

The new constitution placed most of the power in the hands of Parliament. This constitution described the specific roles that the water boards, the provinces, and Rijkswaterstaat would play in water management. Rijkswaterstaat's role was now clearly defined—overseeing public works and executing projects of national importance. In addition to providing technical services, Rijkswaterstaat began to provide advice to the central government on water policy matters.

The second half of the nineteenth century was a time of growth for Rijkswaterstaat. Many successful projects were implemented, including construction of new rivers and canals (the Nieuwe Merwede, the Bergse Maas, the Noordzee Kanaal, and the Nieuwe Waterweg). A private company, under the direction of Rijkswaterstaat, constructed the Noordzee Kanaal. The Nieuwe Waterweg project was completed entirely by Rijkswaterstaat.

The position and authority of the Rijkswaterstaat engineers improved as the training left the military school environment. In 1842 the Royal Academy of

Civil Engineers was established to train members of the Rijkswaterstaat. This institution was later to become the University of Technology in Delft.

By 1900 several major projects were completed, and Rijkswaterstaat had developed into a slow-moving government bureaucracy. The situation was so bad that the Zuiderzee project required the establishment of a separate service. In the 1920s Rijkswaterstaat was additionally criticized for its inability to use new technologies developed in Germany and the United States.

A restructuring of Rijkswaterstaat occurred in 1930. New departments were added to the existing set of regional offices. These new departments were established to provide support for specific projects and techniques (i.e., bridges, sluices, and weirs).

The crowning achievement of Rijkswaterstaat was the management of the Delta Plan established after the 1953 Flood (see Chapter 8). While this project was considered to be a major achievement of Rijkswaterstaat, it also resulted in a great deal of criticism of the agency as environmental concerns surfaced during the construction of the Oosterschelde Barrier.

Today, the primary role of Rijkswaterstaat is still the management of water issues of national concern, while the water boards are responsible for water issues of local and regional concern. Specifically, Rijkswaterstaat is in charge of water management issues in the North Sea, the Waddenzee, the major rivers, and the estuaries. In addition to many engineers, its ranks include biologists, planners, and behavioral experts. Environmental protection and preservation of the cultural landscape now are integrated into any new plans.

Water Management Policy and Institutions

This section describes recent policies for water management in the Netherlands. Recent changes have moved water management policy from a "top-down" approach to one that tries to seek consensus from a wide variety of groups. Ideally, this approach encourages stakeholder involvement, leading to more effective policies. Recent policy changes have also recognized that water quantity issues cannot be resolved without considering other factors such as water quality, the environment, spatial planning, and nature management. The concept of integrated water management recognizes the need for cooperation and coordination between various governmental bodies when trying to solve water management problems.

The earliest public water management policies date back to the twelfth century. These started a long-running tradition of highly decentralized water management in the Netherlands. The main agency implementing water management policy has, for centuries, been the water boards. The constitution of 1848 reinforced the position of the Dutch water boards. Constitutional revisions in 1983 further defined the authority of the water boards, placing them in a constitutional position similar to that of the municipalities (Oolsthoorn and Tol 2001). More recent water management policy acts are slowly moving the country away from this decentralized approach.

Specific laws have been introduced over the years, often in response to flooding events. After the 1953 Flood disaster, the Delta Act was put into place. This act attempted to reduce flood risk by reinforcing the dikes and by closing the southwest delta coastline. This act transformed flood management from a local concern to a national concern.

The Water Management Act of 1989 determined the main competencies in Dutch water management. The central government is responsible for strategic national policy, water legislation, management of national surface waters (coast and large rivers), and supervision of lower agencies. Water management policy and legislation comes out of the Ministry of Transport, Public Works, and Water Management. The operations department within the ministry is the State Water Management Authority or Rijkswaterstaat. The provinces are in charge of provincial strategic water policy and supervision of lower agencies. The Water Management Act initially gave the provinces charge of operational groundwater management, but this is now being handed over to the water boards. The water boards are in charge of operational surface water management, operational groundwater management, and flood risk management within their district. The municipalities are in charge of operation and maintenance of sewage collection systems, while the water boards manage sewage treatment.

The Water Board Act of 1992 reaffirmed that the water board would be the central agency for local and regional water management. This act provides the present framework for financing, participation, and responsibilities.

In response to the river flood events of 1993 and 1995, a commission was established to look further into reducing flood risk. The result was the "Delta Plan Large Rivers" (1995). This plan accelerated the process of river dike reinforcement, and it established a project priority listing. High priority projects were funded under this act (Oolsthoorn and Tol 2001).

The Water Defense Act was approved in 1996. This act funded the lower priority projects from the project priority list (described above). The Water Defense Act also guaranteed a specific level of protection against flooding depending on the location and situation. It introduced the concept of a "dike ring" area. This is a specific region completely surrounded by a "primary embankment" consisting of dikes, dunes, and other water defense structures. Fifty-three dike ring areas were defined, and each was assigned a specific safety norm. This norm is a measure of how well the entire dike ring area is protected from outside water by its system of primary embankments. This act set in place measures to ensure that the desired level of protection is sustained. It required that the water boards report the condition of the primary embankments to the provinces every five years. Rijkswaterstaat was also required to establish technical guidelines regarding the design, maintenance, and assessment of these primary embankments. Because of this act, any new project must now include a project plan that outlines measures to protect the landscape, nature, and cultural heritage values of the region.

The Water Defense Act formally established the safety standards for all dike rings in the Netherlands (TAW 2000). These standards had evolved over many years. The standards were based on a design water level that is associated with a specific frequency storm event. Each dike ring area is required to be able to safely withstand outside water at this level. The storm event recurrence intervals established in this act vary from 1 every 1,250 years to 1 every 10,000 years.

These values were based on several factors. Higher safety standards were established for areas with higher value and higher population density. Areas along the rivers were given somewhat lower safety standards. This was done because flood warnings can be issued several days in advance, providing more time to evacuate, and because freshwater flooding causes less damage than saltwater flooding. Furthermore, far less damage is caused to dikes by river floods than by coastal floods. In some areas the safety standard was reduced further to protect the landscape, nature, and cultural heritage of an area. In other words, physical modifications made to meet higher safety standards often include the strengthening and raising of the dikes. This may subsequently result in the destruction of historic properties near the dike and alteration of the nearby landscape. Lowering the safety standard can protect these areas. The current safety standards are shown in **Figure 9-1**.

The safety standards set by law and shown in **Figure 9-1** do not measure the actual probability of a flood disaster within a given dike ring area for several reasons. First, the dikes were designed with specific factors of safety. For example, a freeboard was specified requiring the top of the dike to be located a specific distance above the design flood elevation. Also, large areas protected by a long dike ring have a higher probability of flooding because one break anywhere can cause a disaster. The safety standards do not account for all modes of dike failure, including those that could occur with water levels below the design level. The designated safety standard also does not account for risks associated with the many hydraulic structures that are integral to a dike ring. Finally, the current safety standards do not account for the dangers associated with climate change or further land subsidence. In other words, the design standards specify criteria for the likelihood of an initiating event, not the likelihood of failure.

Project FLORIS (Flood Risk and Safety in the Netherlands) was initiated in 2000 to determine the probability of flooding for each of the 53 defined dike rings. The methods applied attempted to account for all of the potential dangers to a dike ring, including further climate change and land subsidence. This project will eventually provide a new basis for reestablishing flood safety levels in the Netherlands.

Policy planning is one of the important features of Dutch water management. Rijkswaterstaat coordinates planning at the national level (Oolsthoorn and Tol 2001). Since 1968 there have been four strategic water

Figure 9-1: Current safety standards for dike rings in the Netherlands. Each dike ring must withstand outside water levels exceeded with the given probability.

Redrawn from TAW 2000, Fig. 1.

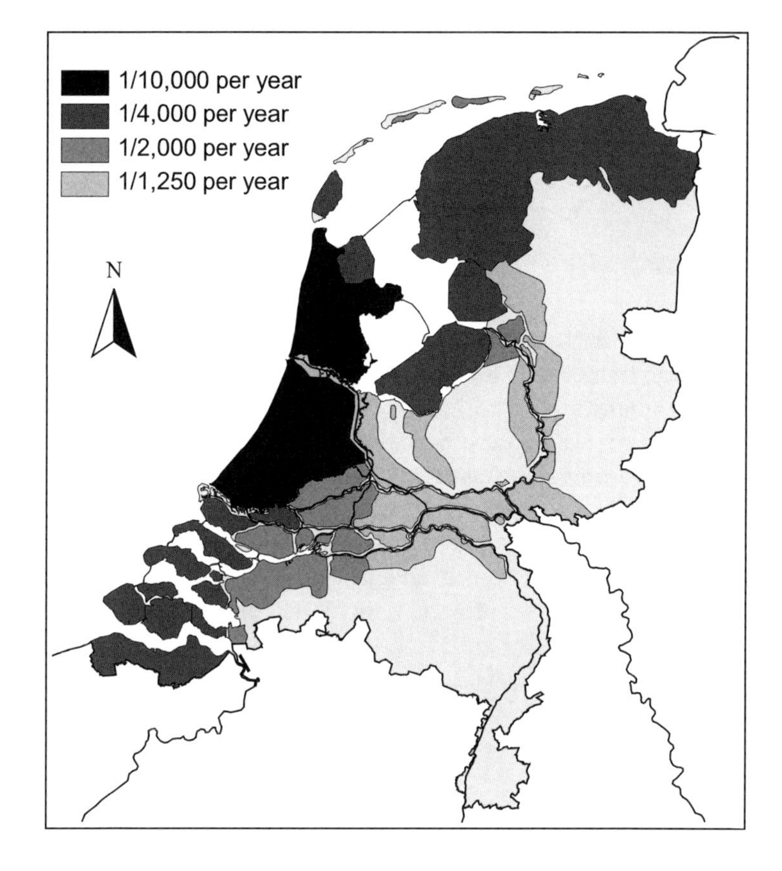

policy plans. The Third Strategic Water Policy Plan established the concept of Integrated Water Management discussed earlier. This integrated approach was widely accepted but is difficult to implement. It requires agencies that acted largely independently in the past to work together.

The current Fourth Strategic Water Policy Plan was presented to parliament in December 1998. It was developed in conjunction with the ministries of Agriculture, Nature Management, and Fisheries, as well as Housing, Spatial Planning, and the Environment. The Association of Water Boards also participated in the process. The primary goal of this plan was to promote a safe and habitable country with sustainable water systems. There was a strong emphasis on flood protection.

Fearing that the current plan did not sufficiently address the problem of future climate change and land subsidence, another commission was given the task of looking to the needs of the twenty-first century. The Committee Water Management 21st Century was established in 1999. It has concluded that in order to ensure a sustainable future, the Netherlands can no longer continue to simply raise the dikes. In 2001 the committee produced a set of guidelines, which were subsequently approved by the Dutch government. The main points were the following:

- **Awareness:** The government must communicate the nature and scope of current flood risks.
- **New safety approaches:** First, officials in charge of water management must anticipate problems instead of just responding to them. Second, water problems must not be shifted to others further downstream. Third, more room must be given to the rivers to increase flood storage and improve conveyance.
- **Storage:** Sites must be found for occasional storage of excess water.
- **Spatial planning:** More efforts must be made to prevent human activities in flood plains. The water boards and municipalities should have a larger role in the spatial planning process. The water boards have a right as well as a responsibility to test all spatial plans against the needs of water management. The municipalities, working together with the water boards, then, have the right to refuse new spatial land use plans under consideration by the central government (RWS-HKW n.d.; van Stokkom and Smits 2002).

With new water management policies based on these guidelines, it is hoped that the Dutch people will be able to keep their feet dry well into the twenty-first century.

10

Excursions to Flood Protection and Land Reclamation Sites

Chapters 2 through 8 list places to visit for those traveling to the Netherlands and wishing to see some of the sites described in this book. This chapter describes six excursions to these places. These excursions satisfy two goals. First, they provide a route between many of the places to visit listed in the book. Second, they provide a way to see some of the unique Dutch landscape along the way.

These excursions require the visitor to travel by automobile. For visitors from the United States, driving in the Netherlands should not be too difficult. The Dutch have built an excellent road system, and Dutch drivers are well trained and generally follow the rules. Dutch highways are often congested, and traffic jams are common. Before renting a car, you should read a travel guide that covers driving in the Netherlands. Familiarize yourself with driving rules and traffic signs used in the Netherlands. The most important advice is to drive alert and follow all the traffic laws—especially speed limits. (The Dutch police use unforgiving cameras to catch speeders.) You should also use a navigator—someone who is not driving but watching the maps and traffic signs.

The Dutch do not use compass directions to indicate which way to turn onto a route. Instead, they use place names. So, for example, when entering the A4 expressway you may have a choice of direction Amsterdam or direction Den Haag. Going in direction Den Haag does not mean that you are actually going to that city, it simply indicates that you are heading on that highway route. This place name direction approach will be used in this chapter.

The limited access divided highways in the Netherlands use an "A" prefix. Secondary highways (often two-lane) use an "N" prefix. Secondary highway extensions of the limited access routes use the same number as the limited access route (i.e., N9 is the secondary highway extension of A9).

An overview of all of the excursion routes is shown in **Figures 10-1** and **10-2**. All excursion routes begin at the A10 ring road around Amsterdam. The listed excursion distances (in kilometers) also start at the A10. A more detailed map than those provided here is necessary to follow these excursion instructions without getting lost. Most excursions will require a national map at a scale of 1:250,000. A better option is to buy one (or more) of the 1:200,000 scale regional ANWB maps—"Noord-Nederland," "Midden-Nederland," or "Zuid-Nederland." (ANWB, Algemene Nederlandsche Wielrijders-Bond, is the Dutch equivalent of AAA.) The best option, which allows further exploration of the Dutch countryside without the danger of getting lost, is to purchase 1:100,000 scale provincial tourist maps ("toeristenkaart"). All of these maps can be purchased at most bookstores or visitor information centers (VVVs). These visitor's centers can be found in most Dutch towns. Look for a sign with three letter "V"s in a triangular arrangement. Occasionally, roads listed on these

Figure 10-1: Northern excursion routes.

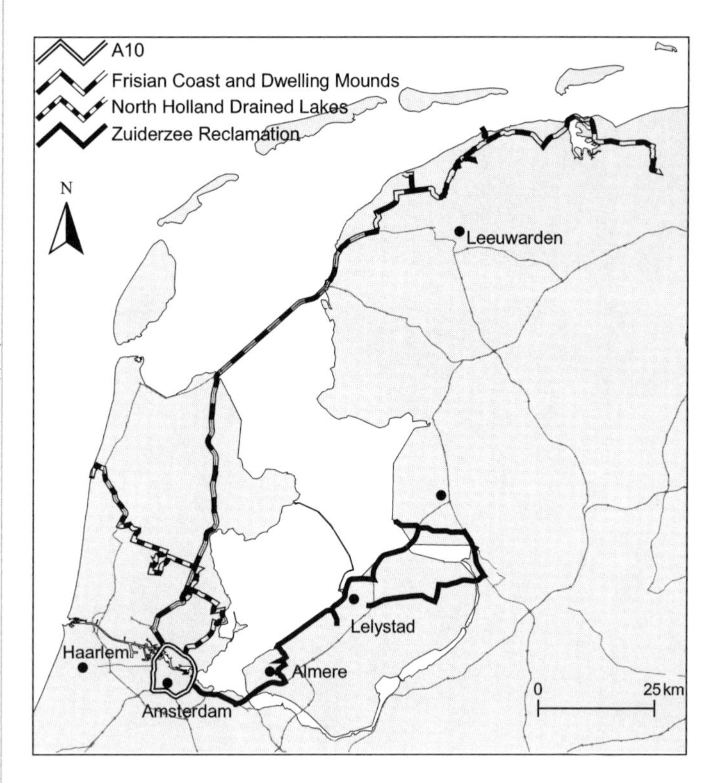

excursion routes may be temporarily closed. A good quality map will allow you to find your way around most detours.

These excursions will take you to many of the sites mentioned in this book. There are many other interesting places to visit along the way. With a good map in hand, you are highly encouraged to wander off the route and explore.

Frisian Coast and Dwelling Mounds

This excursion takes you through some of the wide-open countryside of the northern provinces of Friesland and Groningen. It includes the Zuiderzee Barrier Dam, dwelling mounds, and the Waddenzee coast. It is approximately 225 kilometers (140 miles) long and uses the Friesland and Groningen province maps or the Noord-Nederland regional map.

From the Amsterdam ring (A10) follow A8 then A7 north in the direction Leeuwarden. As you pass the Medemblik exit, you will descend into the Wieringermeer Polder (part of the Zuiderzee reclamation). Note the slight change of elevation and the more regular arrangement of the countryside. Continue on A7 as it takes you to the Barrier Dam (see **Figure 5-11**). Approximately one-fifth of the way across you can stop at the visitor's center (kilometer 70). It includes a restaurant, souvenir shop, and viewing tower. The restaurant displays a large collection of Barrier Dam construction photos.

After crossing the Barrier Dam take N31/A31 in the direction Leeuwarden. Leave A31 at the Franeker exit (kilometer 111). Follow routes N384 and N393 through the towns of Dongjum, Tzummarum, and Minnertsga. Tzummarum is the first dwelling mound town along the route. Note that many of the village names on the road signs in this region are written in both the Dutch and the Frisian languages. Just before Minnertsga look for a house mound on the left and a field mound on the right. Minnertsga was once located on the shore of the former Middelzee (see **Figure 7-7**). There are no dwelling mounds in the former Middelzee because it was drained after dwelling mound construction ended.

Continue on N393 through Sint Jacobiparochie and on to Sint Annaparochie. At the center of Sint Annaparochie (kilometer 126) turn left and follow signs to Nij Altoenae and then to Zwarte Haan. Along this path you will cross several sea dikes that mark the progress of the reclamation of the Middelzee. The first (at Nij Altoenae) is the Oude Bildtdijk, constructed in 1505. As you turn left toward Zwarte Haan, you will be driving on the Nieuwe Bildtdijk, constructed in 1600. At Zwarte Haan (kilometer 132) you can walk onto the present outer dike for a good view of the Waddenzee. You will also see areas that were reclaimed using the Schleswig-Holstein system.

Return to Sint Annaparochie and N393. Drive through Vrouwenparochie and on to Stiens (kilometer 145). Stiens is on the opposite shore of the former Middelzee. Enter Stiens following the slope up the dwelling mound. You will be able to circle the twelfth century church at the center of this dwelling mound village.

Leave Stiens the same way you entered and proceed north on N357 in the direction of Hallum. This route brings you past a number of old coastal dwelling mound towns. As you pass Hijum you will see a mound on your left that has been excavated, leaving only the church and graveyard. Explore Hallum and Marrum on the way to Ferwerd.

Turn right at the first sign leading you into Ferwerd. You will soon see signs for Hogebeintum (or

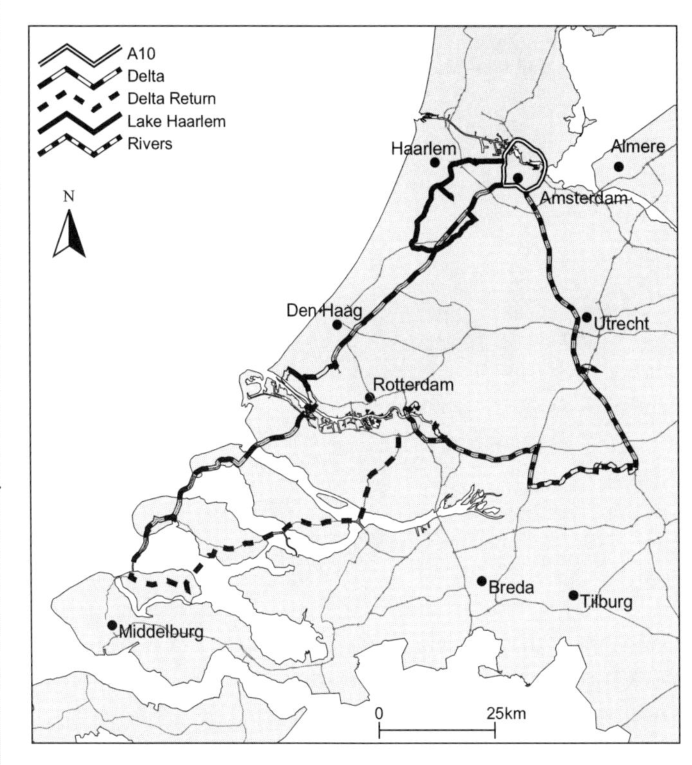

Figure 10-2: Southern excursion routes.

Hegebeintum). As you begin to ascend the Ferwerd mound, turn right toward Hogebeintum. As you approach Hogebeintum, proceed to the "Bezoekers Centrum" or visitor's center (kilometer 158). This is the highest dwelling mound in the Netherlands, rising to the height of 9 meters (30 feet) NAP. Around 1905 much of this village was excavated, leaving the church, graveyard, and a few houses (see **Figure 2-6**). After visiting Hogebeintum, head back to Ferwerd and N357.

At Holwerd turn left on N356 toward the ferry terminal for the island of Ameland. This road takes you over the sea dike and several kilometers into the Waddenzee (kilometer 170). This is a great place to get a better look at this shallow sea.

Return to Holwerd on N356. At Holwerd turn left onto N358 in the direction of Ternaard. If you decide to take a detour into Holwerd, be advised that it is easy to lose your orientation in this little town.

Follow N358 to N361. Turn left on N361 in the direction of Lauwersoog. Past Anjum you head into recently reclaimed sea-bottom land of the Lauwers estuary. This reclamation left a freshwater lake called the Lauwersmeer. You will pass over the outlet sluices for Lauwersmeer (kilometer 197) and on to Lauwersoog. This is the ferry terminal for the island of Schiermonnikoog.

Continue on N361 in the direction of Groningen. At the village of Wehe den Hoorn turn right in the direction of Aduard. After passing the Reitdiep (former Hunze River) and the town of Saaksum, look for signs leading you to the town of Ezinge (kilometer 224). This is the dwelling mound town made famous by archaeological investigations in the early part of the twentieth century. On the road leading up the mound toward the church tower you find the Museum Wierdenland. This is the end of the Frisian Coast and Dwelling Mound excursion.

Noord-Holland Drained Lakes

The goal of this excursion is to see the seventeenth century drained lakes in the part of Noord-Holland north of Amsterdam. You will also see the large sea wall—the Hondsbosse Zeewering—as well as the Noord Hollands Kanaal. This excursion is approximately 88 kilometers (55 miles) long and uses the Noord-Holland province map.

Starting on the Amsterdam ring road (A10), head north on N247 in the direction of Volendam. Just past Monnickendam you will cross a bridge over the Purmer Ee—the outlet of the Purmer ring canal. Turn at the second right past the bridge in the direction of Katwoude. This road will lead along the old Zuiderzee coast to Volendam. There are several places where you can stop and view the surrounding countryside and the former Zuiderzee, now the Markermeer, from the top of this old seawall (see **Figure 3-2**).

Just as you enter Volendam (kilometer 15) turn left (direction Hoorn) on the road that returns you to N247. Continue on N247 in the direction Hoorn. At N244 turn left in the direction of Purmerend. You will immediately descend into the former Purmer Lake, which was drained in the seventeenth century. Follow N247 to the intersection with A7. Head south on A7 (direction Amsterdam) to the next exit—Purmerend (kilometer 28). At the end of the exit ramp turn left toward Purmerend. After driving about 1 kilometer, turn left again at the last opportunity before you cross a canal. (This will be just *after* the road leading to Oosthuizen.)

If you did this correctly, you will be on an extremely narrow road on top of the Beemster Polder ring dike. The ring canal will be on your right and the lower Beemster drained lake will be on your left. Even though this

Figure 10-3: Entering the Beemster drained lake polder.

road is no wider than most bicycle paths, it is a two-way road. You need to watch for oncoming cars and look for places to allow them to get around you.

As you drive along this road, you will get a better sense of how the polder was created (see **Figure 3-3**). Note the regular drainage network within the polder. Note also one of the forts that was once part of the defense line for Amsterdam. Turn left at the street named Oosthuizenweg (kilometer 37). You will now descend into the Beemster Polder at its deepest northeast corner as seen in **Figure 10-3**. Follow this road under the A7, through the town of Noordbeemster, and straight across the polder to the ring dike on the other side. At the ring dike turn left. Follow the dike road until you reach the bridge crossing the ring canal into the town of De Rijp (kilometer 48).

Prior to the seventeenth century, De Rijp was caught in the middle between two large, growing lakes—the Beemster and the Schermer. Without the efforts made in the seventeenth century, this town might not exist today. De Rijp was also the home of the most famous Dutch windmill builder, J. A. Leeghwater. The long main street of De Rijp is very narrow. It requires patience since you often have to wait for opposing traffic to pass by.

Your 1:100,000 scale tourist map should clearly show the difference between the seventeenth century drained polders and the older areas in between. Continue on the road through De Rijp and follow signs leading to Schermerhorn. Before reaching Schermerhorn you will return to another ring dike road. This one follows the Schermer Polder. Just before reaching N243 you will see several windmills. There is a group of three, on your left, at the intersection of the ring dike road and N243 (one of these is shown in **Figure 3-6**). The middle one is the Schermerhorn Windmill Museum (kilometer 57). This museum provides a video presentation in English as well as an opportunity to explore the windmill.

After leaving the Windmill Museum, drive down N243 in the direction of Alkmaar. Take Ring Alkmaar in the direction Den Helder. This will lead to N9, where you continue in the direction Den Helder. Once out of Alkmaar, N9 follows the Noord Hollands Kanaal. This 79-kilometer (49-mile) long, 5-meter (16.5-feet) deep canal was constructed between 1819 and 1825 without the advantage of steam-driven equipment (see **Figure 6-8**).

After passing the exit for Schoorl, you will begin to see the Hondsbosse Zeewering in the distance off to your left. This massive sea wall fills a large gap in the line of coastal dunes. Farther ahead turn left on the road leading to Petten (kilometer 85). When you reach Petten, follow signs leading to Camperduin. This road will follow the Hondsbosse Zeewering (see **Figure 7-5**). There are a couple of observation points where you can park and climb to the top of the sea wall. Here you can get a good view of the Hondsbosse Zeewering and the North Sea. This is the end of the Noord-Holland Drained Lakes excursion.

Haarlemmermeer

The goal of this excursion is to give you a good look at the Haarlemmermeer Polder. This excursion follows the ring dike, includes three museums, and provides a good overview of this area, which was first made dry in the middle of the nineteenth century using steam pumping technology. The excursion is approximately 65 kilometers (40 miles) long and uses the Noord-Holland province map or the Midden-Nederland regional map.

Exit the western portion of the A10 ring road in the direction of Haarlem. This will lead you to N200. Halfway between Amsterdam and Haarlem is the town of Halfweg. The steam pumping museum Stoomgemaal Halfweg is located here. The museum is located in the pumping facility that was used to pump water out of the Haarlemmermeer ring canal and into the IJ. To find this museum, turn left (off N200) at the main intersection at Halfweg. The name of the street is Oranje-Nassaustraat. Turn right at the next intersection—Julianastraat. At the end, turn right on Haarlemmermeerstraat. This should bring you to the museum (kilometer 6).

After a visit at the museum, return to Oranje-Nassaustraat. Since Haarlemmermeerstraat is a one-way street, you will have to drive around the block to get back to Julianastraat. Turn right from Julianastraat onto Oranje-Nassaustraat. Follow this street over the Haarlemmermeer ring canal. Turn right immediately after the canal onto Zwanenburgerdijk. This is the Haarlemmermeer ring dike road. You will follow this road in a counterclockwise direction around this former lake. As you follow this path, you should continue to keep the Haarlemmermeer ring canal immediately on your right hand side. In a few locations where the main road drops into the polder, you should stay on the ring dike.

The next stop is at the Cruquius pumping station (see **Figure 4-15**). This is one of the original three steam pumps used to drain the Haarlemmermeer. To get there, stay on the dike road. The road name changes from Zwanenburgerdijk to Vijfhuizerdijk and then to Cruquiusdijk. The museum is located just before the intersection with N201. The road to the parking lot (kilometer 19) is just before the museum restaurant.

After the Cruquius Museum, the next stop is the Historical Museum Haarlemmermeer located in Hoofddorp (away from the ring dike). Just beyond the Cruquius Museum turn left onto N201 in the direction of Hoofddorp. Follow signs leading you into the park—Haarlemmermeerse Bos. Once in the park follow signs to the museum and the museum parking area (kilometer 24).

After visiting the museum, return to the ring dike at Cruquius to continue the excursion in a counterclockwise direction along the dike road. One of the other original pumping stations was the Leeghwater (kilometer 45). This has now been replaced with a modern pump. This new pumping station is located at the south end of the main north-south drainage canal. To get a better view of the landscape inside the polder, turn left just beyond the pumping station and follow this canal. There is a crossover bridge just below the A44 expressway. Cross to the other side and return to the ring road (kilometer 48).

The ring canal and road pass over the high speed train line (under construction in 2005) and the A4 expressway (kilometer 53). You will also see areas where the ring canal widens significantly. These are parts of the old Haarlemmermeer that were not reclaimed because water storage capacity was still needed for the regional drainage network. Turn left at N201. This will return you to the A4 expressway along the south edge of Schiphol airport. This is the end of the excursion.

Zuiderzee Reclamation

The goal of this excursion is to see the areas reclaimed from the former Zuiderzee in the twentieth century. The excursion goes through the cities of Almere, Lelystad, and Urk. It takes you to the New Land Heritage Centre and the Ramspol storm barrier. It is approximately 157 kilometers (98 miles) long and uses the Flevoland province map.

From the Amsterdam ring (A10) travel east on A1. Exit onto A6 in the direction of Almere. This route passes over the bridge linking the old land with the new polders (kilometer 14). The first stop is at the center of Almere-Stad. This is the central nuclei of this polynuclear city. The land on which Almere sits first came dry in 1968. The first inhabitants settled in Almere-Haven in 1976. Almere-Stad is projected to be finished in 2006 when the population reaches 180,000.

Exit A6 at the Almere-Stad exit. Turn left onto S103 in the direction of Almere Stad. Turn left on Cinemadreef, following signs to the Centrum. This will bring you into the center of Almere Stad. Find a place to park (look for parking route signs) and explore the city center.

Leave the city center via Cinemadreef. At Veluwedreef (S103) turn left. Follow the road as it changes names to Vrijheidsdreef. Turn left at S104 in the direction of Almere-Buiten. Turn left on the ramp leading to S106 Buitenring continuing in the direction of Almere-Buiten. Turn left at the next traffic light onto Grote Vaartweg in the direction De Vaart. This road ends at the polder dike road (kilometer 38) just east of the De Blocq van Kuffeler pumping station. Turn right onto the dike road.

As you travel along this road, the Markermeer will be on your left. A few kilometers ahead you will see the Oostvaardersplassen nature reserve on your right. After following the dike road for about 14 kilometers (8.7 miles), turn right on the road leading to the Oostvaardersplassen nature reserve visitor's center (kilometer 52). Even though you are heading back into the polder, the road is still elevated. This polder was constructed in two phases. This road is on top of the Knardijk—the dike that once was the southwestern boundary of Oostelijk Flevoland. Around 1960 there would have been dry ground on your left and water on your right.

After visiting the nature reserve (kilometer 57), return to the outer dike road. Turn right in the direction of Lelystad. Follow this road along the western edge of the city. The road will move off the dike in a couple of places, but you should not veer away from the dike. You will pass the pump station Wortman (kilometer 67). Beyond this point, look for and follow signs leading you to Batavia Stad. Once in this commercial development, look for signs to the New Land Heritage Centre (Nieuwe Land Erfgoedcentrum) (kilometer 70). This is an excellent museum covering, among other things, the technical aspects of the development of the IJsselmeer polders.

After leaving the museum return to your original route following the western edge of the polder. Very soon you will see N302. To the left, on N302, is the Markerwaard dike road leading to Enkhuizen (kilometer 72). At this point continue straight in the direction of Swifterbant. This will bring you to A6. Head north on A6 in the direction Emmeloord. This takes you over the bridge between the Flevoland Polders and the Noordoost Polder (kilometer 91). The bridge gives you the best view of the bathtub-like configuration of the IJsselmeer Polders.

Take the first exit and turn left on N352 in the direction of Urk. As you drive to this former island, you will notice a distinct difference between the World War II era construction of the polder farmsteads and the much older former island town. On your left, you will also be able to see the dike in the distance, keeping the waters of the IJsselmeer from filling the entire polder. Just as you enter Urk, you will pass one of the Noordoost Polder pump stations (kilometer 101). Park in Urk and walk along the harbor area in the direction of the lighthouse. Just beyond the lighthouse you will find a memorial to the fishermen from Urk who lost their lives on the sea.

Leave Urk the same way you entered. Do not return to the A6, but continue on N352. This will lead you to the other former island, Schokland (kilometer 115). Here there is a small museum and visitor's center.

After leaving Schokland, continue east on N352. Turn right at N50 in the direction of Kampen. Shortly, you will cross the bridge that leads you back to the "old" land at Ramspol. On your right you will see the inflatable storm barrier (see Figure 5-20) (kilometer 121). The road leading to the southern end of the structure appears on your right just beyond the bridge. This may be a good place to stop and walk back to the storm barrier for a better view. This turn also takes you to the visitor's center (open only by appointment and only for groups of ten or more).

Return to N50 in the direction of Kampen. Soon after crossing the IJssel River look for signs leading you back to Lelystad. This route crosses back into the Flevoland Polder. Here you will be able to get a close-up view of the extensive agricultural lands that were developed from the bottom of the former Zuiderzee. Continue to follow signs leading to Lelystad, the end of the excursion (kilometer 156).

Delta Works

The Delta Works excursion brings you to the most impressive structures built as part of the Delta Plan. There are only three stops on this excursion—the Maeslant Storm Surge Barrier, the Haringvliet Discharge Sluices, and the Oosterschelde Storm Surge Barrier. It is approximately 150 kilometers (93 miles) long and uses the Zuid-Holland and Zeeland province maps or the Zuid-Nederland regional map.

Figure 10-4: Support pier, road deck, gate, and lifting mechanism of the Oosterschelde Barrier.

From the A10 ring around Amsterdam follow A4 south in the direction of Den Haag. Continue on A4 past Den Haag, exiting at the Den Hoorn exit. Follow N223 in the direction of Hoek van Holland. Before reaching Hoek van Holland this route will lead you onto the Maasdijk or route N220 (kilometer 64). This dike road gives you a good view of the surrounding Westland greenhouse district. Continue on N220 in the direction of Hoek van Holland until you see signs for Maeslant Kering. The signs will lead you to the left, bringing you to the banks of the Nieuwe Waterweg and eventually to the Maeslant Storm Surge Barrier—one of the greatest engineering achievements in the world (see **Figure 8-27**) (kilometer 71).

After visiting the Maeslant Barrier, return to the Maasdijk and go in the direction of Rotterdam. Follow the signs leading you into the town of Maassluis and then on to the town of Rozenburg by ferry. The ferry boat (kilometer 87) crosses every 20 minutes. The ferry is the best way to get a view of this busy shipping channel. Once off the boat, follow the signs in the direction of Brielle. This will take you first to N15/A15 and then to N57. You will then simply stay on N57 for the rest of this excursion.

The first major structure that you cross will be the Haringvliet sluices (see **Figure 8-16**). Just beyond the structure look for signs for Expo Haringvliet (kilometer 108). After a visit there, return to N57 in the direction Goedereede.

You will not be able to miss the Oosterschelde Barrier (**Figure 8-25**). The hydraulic cylinders that raise and lower the gates can be seen long before you reach the barrier. The barrier consists of three groups of gates. After the second group of gates, look for signs leading you to the visitor's center at Waterland Neeltje Jans (kilometer 149). The photograph in **Figure 10-4** was taken from the section of the structure open to visitors. The top of a support pier is shown in the lower right, the bottom of the road deck is shown in the upper right, and the top of the gate and the lifting mechanism are shown on the left.

From Waterland Neeltje Jans, you can return by the same route you came. An alternative return route runs along some of the inner dams of the Delta Project. The route follows N57 (direction Middelburg), N255, N256, N59, and A29 to Rotterdam and takes you across the Zeeland Bridge, Grevelingen Dam, Volkerak Dam, and Haringvliet Bridge.

Rivers

The Rivers excursion takes you along parts of the branches of the Rhine River in the Netherlands. Here you will see some of the typical river shore landscapes including the river dikes, locks, bridges, and water control structures. You will visit the Visor Weir on the Lek River, the windmills at Kinderdijk, and the storm barrier on the Hollandse IJssel River. This excursion is approximately 165 kilometers (103 miles) long and uses the Utrecht, Gelders Rivierengebied, and Zuid-Holland province maps or the Zuid-Nederland regional map.

From the A10 ring around Amsterdam turn on the A2 in the direction of Utrecht. Continue past Utrecht and follow A2 until you reach the Vianen exit. Follow the road first in the direction of Vianen then Vianen-Oost and Hagestein. Immediately after crossing the drawbridge over the Merwede Kanaal go to the left. When the main road turns right, continue straight, following the canal. This will lead you to a road on the winter dike along the south bank of the Lek River. Follow this to the Visor Wier

Figure 10-5: Kinderdijk windmills.

(see **Figure 6-9**) (kilometer 46). There is a small parking and viewing area. Return to A2 by the same path.

Continue south on A2 in the direction of s'-Hertogenbosch. You will enter the A2 at the intersection with the A27. Watch carefully that you proceed on the correct road. Just beyond the bridge over the Waal River (kilometer 73), exit at Zaltbommel. Follow the street Steenweg into the town of Zaltbommel. This road will rise to the top of the dike. At the top, turn left on the Gamerschedijk (kilometer 79). You are now on the winter dike road on the south bank of the Waal River. To the south of this road is the Bommelerwaard, which was completely reconstructed between 1966 and 1975.

Follow this road to Zuilichem. Turn left on the road Meidijk just beyond Zuilichem (look for signs to Giessen). This will lead you to N322, where you turn right, heading first in the direction of Giessen, then Gorinchem. Along the way you will cross the Afgedamde Maas (kilometer 94). This was the route of the Maas River prior to major modifications made in the later part of the nineteenth century. At A27 head north in the direction of Gorinchem. After crossing the Waal River follow A15 west in the direction of Rotterdam. Leave A15 at the Alblasserdam exit. From this exit you should be able to easily follow the signs leading you to Kinderdijk (kilometer 140). This location, along the Lek River, has the highest concentration of working windmills in the country. Since this is a popular tourist stop, there is ample parking. Here you can walk along a paved path that takes you past many of these windmills as is seen in **Figure 10-5**.

After your visit to Kinderdijk, return to A15. Continue in the direction of Rotterdam. This will require switching to A16 (direction Den Haag). A16 splits into local and express lanes. Get into the local lanes by following signs to Capelle a/d IJssel. After crossing the Nieuwe Maas River take the next exit and get on N219 in the direction of Capelle a/d IJssel. After three kilometers, turn right onto N210 in the direction of Krimpen a/d IJssel. As you cross the Hollandse IJssel River, you cannot miss the storm barrier placed there, the first project of the Delta Plan (kilometer 164). If you continue over the bridge, you can find local roads that lead you back to the dike road. You can park on the dike road to get a better view of this structure and even cross the river on a pedestrian path to get a closer look (see **Figure 8-9**).

These excursions were designed to give visitors to the Netherlands an opportunity to discover for themselves how the Dutch have used technologies over many centuries to keep their feet dry. The problems that the Dutch will face in the future will be no less challenging than those of the past. Future climate change, land subsidence, and sea level rise will force the Dutch to either expand their defenses or find new ways to live with the water. Hopefully, visitors will still be able to follow these excursions well into the future.

Glossary of Dutch Words and Phrases

Amsterdams Peil
national elevation datum established in 1682

bezoekers centrum
visitor's center

binnenkruier
windmill with turning mechanism located inside the structure

boezem
polder discharge storage reservoir

bovenkruier
windmill with rotating caps

buitenkruier
windmill with turning mechanism located outside the structure

diep
deep

dijk
dike

dijkgraaf
chairman of a water board

dobbe
freshwater pond constructed at the center of a dwelling mound

droogmakerijen
drained lake polder

gat
hole

heemraden
member of a water board

hoogheemraden
member of a regional water board

kanaal
canal

keuren
rules for maintenance and operation of drainage/dike systems

meer
lake

nieuw
new

noord
north

noordoost
northeast

Normaal Amsterdams Peil (NAP)
national elevation datum revised in nineteenth century

oost
east

oostelijk
eastern

oud
old

paaldijk
wooden pile dike

plassen
man-made lakes (often a result of peat dredging operations)

polder
water management unit where groundwater levels are controlled

rietdijk
reed dike

Rijkswaterstaat
government agency in charge of water management on a national scale

slikkerdijk
mud dike

stadhouder
provincial executive

terp
dwelling mound in province of Friesland

toeristenkaart
tourist map

Verenigde Oostindische Compagnie
Dutch East Indies Company

VVV
Netherlands Tourist Bureau

waterschappen
water boards or local/regional water authorities

waterweg
waterway

west
west

wierde
dwelling mound in the province of Groningen

wierdijk
weir dike

zee
sea

zuid
south

zuidelijk
southern

Glossary of Dutch Place Names

NAME	PROVINCE	CITY, TOWN, VILLAGE	RIVER	CANAL	LAKE	ISLAND	ESTUARY, INLET, INLAND SEA	DAM, DIKE, SEAWALL, BARRIER	POLDER	NOTES
Achtermeer					■				■	Drained in 1533
Aduard		■								Along Frisian Coast and Dwelling Mounds Excursion
Afgedamde Maas			■							Dammed portion of Maas River
Afsluitdijk								■		"Enclosing Dike" in Zuiderzee reclamation
Alblasserdam		■								Along Rivers Excursion
Alblasserwaard									■	Drainage outlet at Kinderdijk
Alkmaar		■								Along Drained Lakes Excursion
Almere (lake)					■					Became Zuiderzee in twelfth century
Almere (city)		■								Largest city in Zuiderzee reclamation
Almere-Buiten		■								Almere nuclei
Almere-Haven		■								Almere nuclei
Almere-Hout		■								Almere nuclei
Almere-Poort		■								Almere nuclei
Almere-Stad		■								Almere city center

	NAME	PROVINCE	CITY, TOWN, VILLAGE	RIVER	CANAL	LAKE	ISLAND	ESTUARY, INLET, INLAND SEA	DAM, DIKE, SEAWALL, BARRIER	POLDER	NOTES
	Ameland						■				Waddenzee Barrier island
	Amstel			■							Passes through Amsterdam
	Amsteldiep							■			Former Zuiderzee inlet
	Amsteldiep Dike								■		First dam in Zuiderzee Project
	Amsterdam Rijnkanaal				■						Connects Amsterdam and Rhine River
	Andijk									■	Zuiderzee research polder
	Anjum		■								Along Frisian Coast and Dwelling Mounds Excursion
	Arkelse Dam								■		Location of steam-driven scoop wheel
B	Beemster					■				■	Drained in 1612
	Bergermeer					■				■	Drained in 1564
	Bergse Maas			■							Reconstructed section of Maas River
	Biddinghuizen		■								Constructed in Zuiderzee reclamation
	Biesbosch										"Reed forest" in flooded Grote Waard
	Blijdorp Polder									■	Early steam drainage attempt
	Blokzijl		■								Terminus of Noordoost Polder Dike
	Bommelerwaard									■	Consolidated river polder
	Boorne			■							Drained to former Middelzee
	Braakman							■			Formed in 1373 Flood, closed in 1952
	Brielle		■								Along Delta Works Excursion
	Brielse Maas			■							Portion of Maas River near Brielle
	Brielse Meer					■					Formed from damming Brielse Maas
	Brouwers Dam								■		Delta Plan dam
	Brouwershavense Gat							■			Southwest delta sea inlet
C	Cabauw		■								Location of hollow post mill
	Camperduin		■								South end of Hondsbosse Zeewering
	Capelle a/d IJssel		■								Along Rivers Excursion
D	De Rijp		■								Home of Jan Adriaensz Leeghwater
	De Vaart										Almere industrial zone
	Delf				■						Built in twelfth century
	Delfzijl		■								Terminus of Delf Canal
	Den Haag		■								Known in English as "The Hague"

NAME	PROVINCE	CITY, TOWN, VILLAGE	RIVER	CANAL	LAKE	ISLAND	ESTUARY, INLET, INLAND SEA	DAM, DIKE, SEAWALL, BARRIER	POLDER	NOTES
Den Helder		■								Terminus of Noord Hollands Kanaal
Den Oever		■								Leemans pumping station location
Dergmeer					■				■	Drained in 1542
Diefdijk								■		River polder cross dike
Diemerdijk								■		Zuiderzee coast pile dike
Dollard							■			Formed in 1362
Dongjum		■								Along Frisian Coast and Dwelling Mounds Excursion
Dordrecht		■								Grote Waard town not flooded in 1421
Drenthe	■									Northeast Province
Dronten		■								Constructed in Zuiderzee reclamation
Drontermeer					■					Zuiderzee reclamation marginal lake
Eemmeer					■					Zuiderzee reclamation marginal lake
Eems							■			Estuary in northeast
Eenum		■								Excavated dwelling mound
Egmondermeer					■				■	Drained in 1564
Emmeloord		■								Constructed in Zuiderzee reclamation
Enkhuizen		■								Former Zuiderzee coast city and terminus of Markerwaard dike road
Ezinge		■								Dwelling mound
Ferwerd		■								Dwelling mound
Fivel			■							Early Groningen river
Flevoland	■									Province consisting wholly of Zuiderzee reclamation polders
Franeker		■								Along Frisian Coast and Dwelling Mounds Excursion
Friesland	■									Northwest Province
Gelderland	■									Eastern Province
Giessen		■								Along Rivers Excursion
Godlinze		■								Dwelling mound
Goedereede		■								Along Delta Works Excursion
Goeree						■				Southwest delta island
Gooimeer					■					Zuiderzee reclamation marginal lake

E

F

G

	NAME	PROVINCE	CITY, TOWN, VILLAGE	RIVER	CANAL	LAKE	ISLAND	ESTUARY, INLET, INLAND SEA	DAM, DIKE, SEAWALL, BARRIER	POLDER	NOTES
	Gorinchem		■								Location of Maas and Waal confluence
	Gouda		■								Early steam pump location
	Grevelingen							■			Southwest delta estuary
	Grevelingen Dam								■		Delta Plan dam
	Grevelingenmeer					■					Lake formed from Grevelingen estuary
	Groningen	■	■								Northern province and city
	Grote Waard									■	River polder lost in 1421 flood
H	Haarlem		■								City along former Haarlemmermeer
	Haarlemmermeer					■				■	"Lake Haarlem" drained in 1852
	Hagestein		■								Along Rivers Excursion
	Halfweg		■								Haarlemmermeer ring canal outlet
	Hallum		■								Dwelling mound
	Haringvliet							■			Southwest delta estuary
	Haringvliet Brug										Bridge across Haringvliet estuary
	Haringvliet Dam								■		Delta Plan dam
	Heemstede		■								Early steam pump location
	Heerewaarden		■								Between Maas and Waal Rivers
	Heerhugowaard					■				■	Drained in 1625
	Hegebeintum		■								Frisian spelling of Hogebeintum
	Hellevoetsluis		■								Early steam pump location
	Hijum		■								Dwelling mound
	Hoek van Holland		■								North Sea end of Nieuwe Waterweg
	Hogebeintum		■								Highest dwelling mound
	Hollands Diep							■			Southwest delta estuary
	Hollandse IJssel			■							Location of first Delta Plan barrier
	Hollandse IJssel Barrier								■		First barrier built in Delta Plan
	Holwerd		■								Dwelling mound
	Hondsbosse Zeewering								■		Sea wall along Noord-Holland coast
	Hoofddorp		■								Constructed in former Haarlemmermeer
	Hoorn		■								Former Zuiderzee coast city
	Hunze			■							Groningen River
I	IJ							■			Extension of Zuiderzee near Amsterdam

NAME	PROVINCE	CITY, TOWN, VILLAGE	RIVER	CANAL	LAKE	ISLAND	ESTUARY, INLET, INLAND SEA	DAM, DIKE, SEAWALL, BARRIER	POLDER	NOTES
IJssel			■							Primary source of IJsselmeer water
IJsselmeer					■					"Lake IJssel" created from former Zuiderzee
Kampen		■								Along Zuiderzee Reclamation Excursion
Kanaal van St. Andries				■						"St. Andries Canal" linking the Maas and Waal Rivers
Kantens		■								Dwelling mound
Katwijk		■								Located at Old Rhine River mouth
Katwoude		■								Along Drained Lakes Excursion
Ketelmeer					■					Lake between two Zuiderzee polders
Kinderdijk		■								Location with many working windmills
Knardijk								■		Separates Oostelijk and Zuidelijk Flevoland Polders
Kreileroord		■								Constructed in Zuiderzee reclamation
Krimpen aan de IJssel		■								Location where Maeslant Barrier sections were constructed
Krimpenerwaard									■	Location of early steam pump
Kromme Rijn			■							Rhine branch dammed in 1122
Kuinre		■								Former Zuiderzee coast town
Lauwers							■			Northern estuary closed in 1969
Lauwersmeer					■					Formed by damming the Lauwers estuary
Lauwersoog		■								Along Frisian Coast and Dwelling Mounds Excursion
Leermens		■								Dwelling mound
Leeuwarden		■								Capital of Friesland Province
Leiden		■								Formerly on Haarlemmermeer shore
Lek			■							Branch of the Rhine River
Lelystad		■								Constructed in Zuiderzee reclamation and capital of Flevoland Province
Lemmer		■								Terminus of Noordoost Polder Dike
Linge			■							Location of early steam pump
Lith		■								Along Maas River
Lobith		■								Along Rhine River
Loppersum		■								Dwelling mound

	NAME	PROVINCE	CITY, TOWN, VILLAGE	RIVER	CANAL	LAKE	ISLAND	ESTUARY, INLET, INLAND SEA	DAM, DIKE, SEAWALL, BARRIER	POLDER	NOTES
M	Maasdijk								■		Along Delta Works Excursion
	Maasmond							■			Mouth of the Maas River
	Maassluis		■								Along Delta Works Excursion
	Maasvlakte										Land created to expand Rotterdam port
	Maeslant Kering								■		Delta Plan Barrier
	Marken						■				Former Zuiderzee island
	Markerwaard									■	Zuiderzee Polder never constructed
	Marrum		■								Dwelling mound
	Medemblik		■								Terminus of Wieringermeer Polder Dike
	Merwede			■							Starts at confluence of Maas and Waal Rivers
	Merwede Kanaal				■						Connects the Merwede River to the Amsterdam-Rijnkanaal
	Middelburg		■								Capital of Zeeland Province
	Middelstum		■								Dwelling mound
	Middelzee							■			Completely endiked estuary
	Middenmeer		■								Constructed in Zuiderzee reclamation
	Mijdrechtse									■	Early steam pump location
	Minnertsga		■								Along Frisian Coast and Dwelling Mounds Excursion
	Moerdijk		■								Town near the Biesbosch
	Monnickendam		■								Along Drained Lakes Excursion
	Muiden		■								Terminus of Diemerdijk
N	Neder Rijn			■							"Lower Rhine" River
	Nederwaard									■	River polder with Kinderdijk outlet
	Neeltje Jans						■				Terminus of Oosterschelde barrier section
	Nieuw Loosdrecht		■								Near former peat dome
	Nieuwe Bildtdijk								■		Middelzee Dike
	Nieuwe Maas			■							Reconstructed Maas River through Rotterdam
	Nieuwe Merwede			■							Reconstructed branch of Merwede River
	Nieuwe Waterweg				■						"New Waterway" canal completed 1896 between Rotterdam and the North Sea
	Nij Altoenae		■								Along Frisian Coast and Dwelling Mounds Excursion

NAME	PROVINCE	CITY, TOWN, VILLAGE	RIVER	CANAL	LAKE	ISLAND	ESTUARY, INLET, INLAND SEA	DAM, DIKE, SEAWALL, BARRIER	POLDER	NOTES
Nijmegen		■								Along Waal River
Noord Beveland						■				Southwest delta island
Noorderleegdijk								■		Middelzee Dike
Noord-Holland	■									Western Province
Noord Hollands Kanaal				■						Connects Amsterdam and Den Helder, completed in 1824
Noordland						■				Terminus of Oosterschelde Barrier section
Noordoost Polder									■	"Northeast" polder in Zuiderzee reclamation
Noordzee Kanaal				■						Connects Amsterdam with the North Sea, completed in 1876
Nuldernauw					■					Zuiderzee reclamation marginal lake
Oesterdam								■		Delta Plan dam
Ooijpolder									■	Proposed flood storage area
Oostelijk Flevoland									■	"Eastern" Flevoland in Zuiderzee reclamation
Oostergo										Frisian region along Middelzee
Oosterschelde							■			Southwest delta estuary
Oosterschelde Barrier								■		Delta Plan barrier
Oosthuizen		■								Along Drained Lakes Excursion
Oostvaardersdiep							■			Section of former Zuiderzee
Oostvaardersdijk								■		Flevoland Polder Dike
Oostvaardersplassen										Nature area in Zuiderzee reclamation
Oude Bildtdijk								■		Middelzee Dike
Oude Maas			■							"Old Maas" River
Oude Rijn			■							"Old Rhine" River
Ouderkerk		■								Location of river dike break in 1953
Ouwerkerk		■								Location of dike break in 1953
Overijssel	■									Eastern Province
Overwaard									■	River polder with Kinderdijk outlet
Pannerdensch Kanaal				■						"Pannerdensch Canal" constructed to fix Rhine flow distribution
Petten		■								North end Hondsbosse Zeewering
Philips Dam								■		Delta Plan dam

	NAME	PROVINCE	CITY, TOWN, VILLAGE	RIVER	CANAL	LAKE	ISLAND	ESTUARY, INLET, INLAND SEA	DAM, DIKE, SEAWALL, BARRIER	POLDER	NOTES
	Poldijk								■		Middelzee Dike
	Purmer					■				■	Drained in 1622
	Purmer Ee										Purmer polder storage reservoir
	Purmerend		■								Along Drained Lakes Excursion
R	Rammekens										Location of dike breach in World War II
	Ramspol										Location of inflatable dam
	Reimerswaal		■								Destroyed in 1523 Flood
	Reitdiep				■						Along Frisian Coast and Dwelling Mounds Excursion
	Riederwaard										Area flooded in 1373 Flood
	Rijn			■							"Rhine" River
	Rijnmond							■			Old mouth of Rhine River
	Rijnstrangen									■	Proposed flood storage area
	Roggenplaat						■				Terminus of Oosterschelde Barrier section
	Rotte			■							River dammed at Rotterdam
	Rotterdam		■								Port city and site of early steam pump
	Rottum		■								Dwelling mound
	Rozenburg		■								Along Delta Works Excursion
S	s'-Hertogenbosch		■								Noord Brabant city
	Saaksum		■								Dwelling mound
	Schelde			■							Drains to southwest delta
	Schelde-Rijnkanaal				■						Shipping route between Antwerp and Rotterdam
	Schelphoek										Location of 1953 dike burst
	Schermer					■				■	Drained in 1635
	Schermerhorn		■								Along Drained Lakes Excursion
	Scheveningen		■								North Sea beach town
	Schiedam		■								Location where Maeslant Barrier sections were constructed
	Schiermonnikoog						■				Waddenzee Barrier island
	Schiphol										International airport built on lake bottom
	Schokland						■				Incorporated into Noordoost Polder
	Schoorl		■								Along Drained Lakes Excursion

NAME	PROVINCE	CITY, TOWN, VILLAGE	RIVER	CANAL	LAKE	ISLAND	ESTUARY, INLET, INLAND SEA	DAM, DIKE, SEAWALL, BARRIER	POLDER	NOTES
Schouwen-Duiveland						■				Southwest delta island
Sint Annaparochie		■								Along Frisian Coast and Dwelling Mounds Excursion
Sint Jacobiparochie		■								Along Frisian Coast and Dwelling Mounds Excursion
Slootdorp		■								Constructed in Zuiderzee reclamation
Spaarndam		■								Haarlemmermeer sluice outlet
Starnmeer					■				■	First windmill to use Archimedes screw
Stedum		■								Dwelling mound
Stiens		■								Dwelling mound
Swifterbant		■								Constructed in Zuiderzee reclamation
Ternaard		■								Along Frisian Coast and Dwelling Mounds Excursion
Tzummarum		■								Along Frisian Coast and Dwelling Mounds Excursion
Urk		■				■				Incorporated into Noordoost Polder
Usquert		■								Dwelling mound
Utrecht	■	■								Central province and capital city
Valkenburg		■								Roman refuge mound location
Vecht			■							Flows from Utrecht to the IJsselmeer
Veere		■								Near first Delta Plan dam
Veerse Gat Dam								■		First Delta Plan dam
Veluwemeer					■					Zuiderzee reclamation marginal lake
Vianen		■								Along Rivers Excursion
Vlie							■			Early sea inlet
Volendam		■								Along Drained Lakes Excursion
Volkerak							■			Southwest delta estuary
Volkerak Dam								■		Delta Plan dam
Vollenhovermeer					■					Zuiderzee reclamation marginal lake
Vrouwenparochie		■								Along Frisian Coast and Dwelling Mounds Excursion
Waal			■							Main Dutch branch of the Rhine
Waddenzee							■			"Wadden Sea" along northern coast
Walcheren						■				Southwest delta island

NAME	PROVINCE	CITY, TOWN, VILLAGE	RIVER	CANAL	LAKE	ISLAND	ESTUARY, INLET, INLAND SEA	DAM, DIKE, SEAWALL, BARRIER	POLDER	NOTES
Walsweer		■								Dwelling mound
Weerwater					■					Constructed in Zuiderzee reclamation
Wehe den Hoorn		■								Along Frisian Coast and Dwelling Mounds Excursion
Westeremden		■								Dwelling mound
Westergo										Frisian region along Middelzee
Westerschelde							■			Southwest delta estuary
Westerwijtwerd		■								Dwelling mound
West-Friesland Sea Dike								■		Early ring dike
Wieringen						■				Terminus of Wieringermeer Polder Dike
Wieringermeer									■	Zuiderzee reclamation polder
Wieringerwerf		■								Constructed in Zuiderzee reclamation
Wijde Wormer					■				■	Drained in 1626
Winsum		■								Terminus of Delf Canal
Wolderwijd					■					Zuiderzee reclamation marginal lake
Z										
Zaltbommel		■								Along Rivers Excursion
Zandkreek							■			Southwest delta tidal inlet
Zandkreek Dam								■		Delta Plan dam
Zeeland	■									Southwest delta area province
Zeeland Brug										Bridge over Oosterschelde estuary
Zijpe							■			Early sea inlet
Zuid Beveland						■				Southwest delta island
Zuidelijk Flevoland									■	"Southern" Flevoland in Zuiderzee reclamation
Zuiderzee							■			"Southern Sea" reclaimed in the twentieth century
Zuid-Holland	■									"South Holland" Province
Zuidplas Polder									■	Reclaimed with both windmills and steam pumps
Zuilichem		■								Along Rivers Excursion
Zuyderzee							■			Zuiderzee alternative spelling
Zwammerdam		■								Location of thirteenth century dam on Rhine River
Zwarte Haan										Drainage outlet of former Middelzee
Zwartemeer					■					Zuiderzee reclamation marginal lake

References

Boersma, J. W., comp. (1972). *Terpen mens en milieu,* Knoop and Niemeijer, Haren, Netherlands.

Colom, J. A. (1635). *De Vyerighe Colom,* De Vyerighe Colom, Amsterdam.

Dijksterhuis, E. J. (1970). *Simon Stevin: Science in the Netherlands around 1600,* Martinus Nijhoff, Den Haag, Netherlands.

Duin, R. H. A. van, and Kaste, G. de. (1990). *The pocket guide to the Zuyder Zee Project,* [Rijkswaterstaat– Directorate Flevoland], Lelystad, Netherlands.

Elema, A., Klugkist, J. G., and Reinders, C. G., eds. and comps. (1983). *Boerderijenboek Middelstum—Kantens,* Landbouwvereniging Middelstum–Kantens, Kantens, Netherlands.

Gevers, D. T. (1852). *Over de droogmaking van de Haarlemmermeer,* van Cleef and Muller, Leiden.

Harte, J. H. (1849). *Volledig Molenboek,* A. vander Mast, Gorinchem, Netherlands.

Horst, A. Q. C. van der. (n.d.). *Design of the storm surge barrier Ramspol, The Netherlands,* Hollandsche Beton–en Waterbouw bv, Goude, Netherlands.

Hupkes, E. (n.d.). *Transformation of Rotterdam Docklands,* Port of Rotterdam, Rotterdam.

Information and Documentation Centre for the Geography of the Netherlands (IDG). (1994). *Compact geography of the Netherlands,* The Information and Documentation Centre for the Geography of the Netherlands, Utrecht, Netherlands.

Kaijser, A. (2002). "System building from below: Institutional change in Dutch water control systems." *Technology and Culture,* 43(July), 521–548.

Lambert, A. M. (1971). *The making of the Dutch Landscape,* Seminar Press, London.

Lingsma, J. S. (1966). *Holland and the Delta Plan,* Nijgh & Ditmar, Rotterdam.

Lintsen, H. (2002). "Two centuries of central water management in the Netherlands." *Technology and Culture,* 43(July), 549–568.

Meijer, H. (1981). *The region of the Great Rivers, IDG-Bulletin 1980/81,* The Information and Documentation Centre for the Geography of the Netherlands, Utrecht.

———. (1992). *The IJsselmeerpolders, IDG-Bulletin 1992,* The Information and Documentation Centre for the Geography of the Netherlands, Utrecht.

———. (1993). *The South-west Netherlands, IDG-Bulletin 1993,* The Information and Documentation Centre for the Geography of the Netherlands, Utrecht.

———. (1996). *Water in, around and under the Netherlands, IDG-Bulletin 1995/96,* The Information and Documentation Centre for the Geography of the Netherlands, Utrecht.

———. (1999). *The Netherlands 1964–1999, IDG-Newsletter 1999,* The Information and Documentation Centre for the Geography of the Netherlands, Utrecht.

Netherlands Hydrologic Society (NHV). (1998). *Water in the Netherlands,* Netherlands Hydrologic Society, Delft, Netherlands.

Olsthoorn, A. A., and Tol, R. S. J. (2001). *Floods, flood management and climate change in the Netherlands*, Institute for Environmental Studies Vrije Universiteit, Amsterdam.

Oosterschelde Stormvloedkering Bouwcombinatie (DOSBOUW). (1983). *The storm surge barrier in the Eastern Scheldt: For safety and environment*, Oosterschelde Stormvloedkering Bouwcombinatie, Den Haag, Netherlands.

Pols, K. van der, and Verbruggen, J. A. (1996). *Stoombemaling in Nederland, steam drainage in the Netherlands, 1770–1870*, Delftse Universitaire Pers, Delft, Netherlands.

Rijkswaterstaat–Directorate Zuid Holland (RWS-DZH). (n.d.) *Haringvliet in brief*, Rijkswaterstaat–Directorate Zuid Holland, Rotterdam.

Rijkswaterstaat–Directorate Zuid Holland, Directorate Oost Nederland, and the Institute for Inland Water Management and Waste Water Treatment (RWS-DZH, DON,RIZA). (2001). *Room for the Rhine in the Netherlands*, Rijkswaterstaat–Directorates Zuid Holland, Directorate Oost Nederland, and the Institute for Inland Water Management and Waste Water Treatment, Delft, Netherlands.

Rijkswaterstaat–Head Office (RWS-HKW). (n.d.). *A different approach to water, Water management policy in the 21st century*, Rijkswaterstaat–Head Office, Den Haag, Netherlands.

Ritzema, H.P., ed. (1994). *Drainage principles and applications*, 2nd ed., International Institute for Land Reclamation and Improvement, Wageningen, Netherlands.

Spier, P. (1969). *Of dikes and windmills*, Doubleday and Co., New York.

Stichting Wetenschappelijke Atlas van Nederland (SWAVN). (1984). *Bewoningsgeschiedenis*, Atlas van Nederland, no. 2, Stichting Wetenschappelijke Atlas van Nederland, Den Haag, Netherlands.

———. (1986). *Water*, Atlas van Nederland, no. 15, Stichting Wetenschappelijke Atlas van Nederland, Den Haag, Netherlands.

———. (1987). *Bodem*, Atlas van Nederland, no. 14, Stichting Wetenschappelijke Atlas van Nederland, Den Haag, Netherlands.

Stokhuyzen, F. (1962). *The Dutch Windmill*, C. A. J. van Dishoeck, Bussum, Netherlands.

Stokkom, H. T. C. van, and Smits, A. J. M. (2002). Keynote lecture: "Flood Defense in the Netherlands: A new era, a new approach." In *Flood Defense 2000*, Science Press Ltd., New York.

TeBrake, W. H. (2002). "Taming the Waterwolf: Hydraulic engineering and water management in the Netherlands during the Middle Ages." *Technology and Culture*, 43(July), 475–499.

Technical Advisory Committee for Flood Defense in the Netherlands (TAW). (2000). *From probability of exceedance to probability of flooding: Towards a new safety approach*, Rijkswaterstaat–Road and Hydraulic Engineering Division, Delft, Netherlands.

Trilateral Working Group on Coastal Protection and Sea Level Rise (CPSL). (2001). *Coastal protection and sea level rise, Wadden Sea Ecosystem No. 13*, Common Wadden Sea Secretariat (CWSS), Wilhelmshaven, Germany.

Veen, J. van. (1955). *Dredge drain reclaim: The art of a nation*, Martinus Nijhoff, Den Haag, Netherlands.

Ven, G. P. van de, ed. (1994). *Man-made lowlands: History of water management and land reclamation in the Netherlands*, 2nd ed., Matrijs, Utrecht, Netherlands.

———. (2004). *Man-made lowlands: History of water management and land reclamation in the Netherlands*, 4th ed., Matrijs, Utrecht, Netherlands.

Vollmer, M., Guldberg, M., Maluck, M., Marrewijk, D. van, and Schlicksbier, G., eds. (2001). *LANCEWAD: Landscape and cultural heritage in the Wadden Sea region, Project report, Wadden Sea Ecosystem No. 12*, Common Wadden Sea Secretariat (CWSS), Wilhelmshaven, Germany.

Wal, L. T. van der, ed. (1922). *Geschiedenis van de plannen tot afsluiting en droogmaking van de Zuiderzee, met chronologische lijst van geschriften tot en met het jaar 1922*, Algemeene Landsdrukkerij, Den Haag.

Wigbels, V. (n.d.). *Oostvaardersplassen: New nature below sealevel*, MMI Staatsbosbeheer Flevoland-Overijssel, Zwolle, Netherlands.

Index

E

F

G

H

I

J

R

S

T

U

V

W

Z